A Tutorial for Watch Designers

Gerd Plange

A Tutorial for Watch Designers

Springer

Gerd Plange
Beringen, Switzerland

ISBN 978-3-032-33721-4 ISBN 978-3-032-33722-1 (eBook)
https://doi.org/10.1007/978-3-032-33722-1

This Springer imprint is published by the registered company Springer Nature Switzerland AG
The registered company address is: Gewerbestrasse 11, 6330 Cham, Switzerland

Preface

This book, written by a passionate watch designer, is for everyone who loves watches. This includes young people who aspire to a career as watch designers, for whom the book provides a solid foundation for building the necessary knowledge. It also appeals to watch enthusiasts who want to learn more about their objects of desire and how they are designed.

To design a watch is to enter into a dialogue with time itself. A wristwatch is far more than a device for measuring hours and minutes: it is an object that fuses precision with artistry, and practicality with emotion. Every successful design rests on three inseparable pillars—form, function, and emotion—which together shape the timepiece's identity.

Form is the visual language of a watch. The curves of the case, the proportions of the dial, the rhythm of the indices, and the tactile feel of the strap or bracelet all create an esthetic experience. Form speaks before function is ever tested. It is the reason a watch draws attention or feels naturally balanced on the wrist. Good form communicates clarity and elegance, but also character—whether minimalist, bold, or timelessly classical.

Function is the watch's promise to serve reliably. Beyond simply telling the time, the functional dimension considers legibility, ergonomics, durability, and technical performance. From the crisp contrast of the hands against the dial to the reassuring smoothness of a crown being wound, function ensures that the watch fulfils its purpose in daily life. Even the most beautiful design loses its relevance if it fails to perform with consistency and ease.

Emotion is what transforms form and function into something enduring. A watch becomes personal when it tells more than the hour—it tells a story. Perhaps it marks a milestone, reflects a personal style, or connects its wearer to heritage and craft. Emotion is what makes a watch more than an instrument; it becomes a companion or even a legacy passed across generations.

To understand watch design, one must learn to balance these three dimensions. A design too focused on form risks becoming a fragile ornament. One who is too obsessed with function risks sterility. And one that appeals only to emotion may lack clarity and discipline. The art lies in harmonizing all three, so that the watch does not just measure time—it expresses it.

This book consists of four chapters. Chapter 1 contains the basic knowledge a watch designer should possess. Chapter 2 describes the range of design options when designing a watch. Chapter 3 outlines the typical process for designing a watch. Chapter 4 contains tutorials on sketching wristwatches and on creating two- and three-dimensional computer renderings. Some tips on how to take photos of wristwatches round off this chapter.

For those who feel called to become watch designers after reading this book, there are some exercises at the end of each chapter. Solving these exercises is intended to solidify the acquired knowledge and train skills.

The brand "Duvallier" shown on some of the watches and drawings in this book is purely fictitious.

Beringen, Switzerland Gerd Plange

Contents

Chapter 1
What a Watch Designer Should Know

Abstract This chapter provides a comprehensive overview of wristwatch design, beginning with the fundamental components and operating principles of mechanical, tuning-fork, quartz, and hybrid movements. It then classifies watches according to watch-face design, complications, style, specific purpose, target groups, and historical design eras. Particular attention is given to functional and specialized timepieces, including divers', pilots', railroad, racing, medical, and sailing watches. The chapter also examines how aesthetics, technology, functionality, ergonomics, craftsmanship, and historical influences shape watch design, demonstrating the diversity of wristwatches as both precision instruments and expressions of personal style.

The design process is characterized by a sequence of decisions. The more knowledge a designer has about the product being designed, the easier it is for them to make the right decisions regarding specific design features. This first part of the book is therefore dedicated to building a foundation of knowledge about wristwatches.

The "hard facts" of wristwatches are imparted, including their classification and technical features. The "soft facts" concern the character, target groups, and art historical style of wristwatches.

General watch knowledge (that some watch enthusiasts use to impress at parties) relates to watchmaking history, the most important watch designers, and the storytelling of the watch industry. For those seriously interested in watch design, the chapter concludes with information on luxury design and targeted training to become a watch designer.

1.1 Components of a Wristwatch

Mechanical wristwatches are fascinating not only because they measure time without electronics, but also because of the number of components hidden inside their movements. At the extremes, some watches are deliberately designed with very few parts,

G. Plange, *A Tutorial for Watch Designers*,
https://doi.org/10.1007/978-3-032-33722-1_1

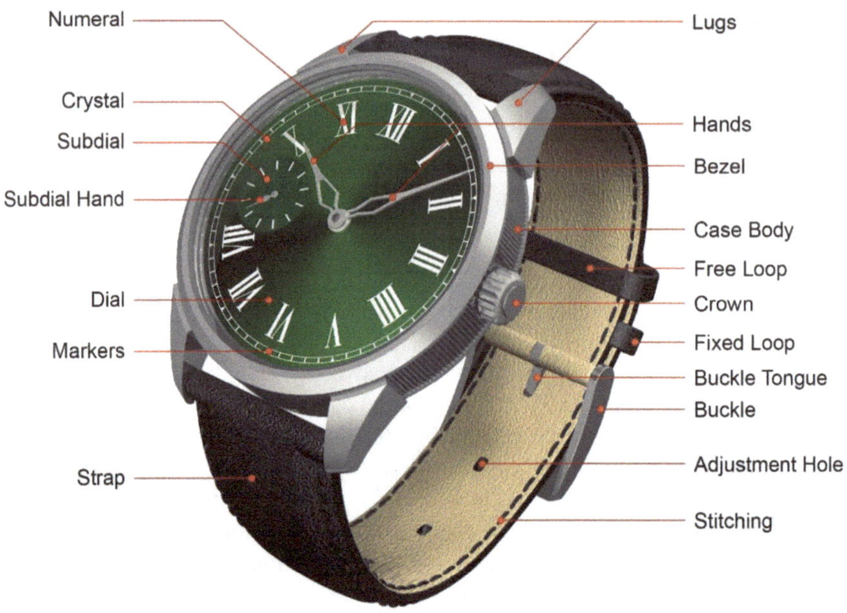

Fig. 1.1 Components of a Wristwatch

while others push engineering to its limits with hundreds or even thousands of components.

Watches with a minimal number of components aim for simplicity, robustness, and efficiency. Fewer parts generally mean fewer friction points, easier servicing, and often greater reliability. A well-known example is the hand-wound mechanical movement with only hours, minutes, and seconds. Such movements can have around 50–80 parts.

At the opposite extreme are high-complication watches, often created by luxury manufacturers and independent watchmakers as technical showcases. These watches can contain several hundred to over a thousand individual components.

Figure 1.1 illustrates a fantasy watch. In its simplicity, it corresponds to the archetype of a wristwatch. It will demonstrate the components that most wristwatches have [1].

1.2 Classification of Watches

Since the invention of the wristwatch, numerous variations have been developed. The following explanations are intended to provide an overview. For this purpose, all wristwatches have been classified according to various characteristics. There are certainly subgroups in the watch market that are not mentioned here. In any case, the most important groups are represented.

1.2.1 Types of Watches by Movement

The "movement" (also called the "calibre") is the mechanism inside the watch that drives the hands and complications, essentially the "engine" of the watch. It is the core system that makes the watch function, powering timekeeping and any additional features [2].

1.2.1.1 Mechanical Movements

A mechanical wristwatch movement is a tiny but complex machine that measures time using purely mechanical components, without electronics or batteries. As shown in Fig. 1.2, its operation can be broken down into several main parts and processes:

Energy Source—The Mainspring

The mainspring is a coiled spring that stores energy when wound, either manually (by turning the crown) or automatically (through a rotor that moves with the wearer's wrist). This spring slowly unwinds, releasing energy to power the watch.

Gear Train—Transmitting the Energy

The energy from the mainspring is transferred through a series of gears called the gear train. These gears reduce the speed of the unwinding spring and distribute power to different parts of the movement, including the hands of the watch.

Escapement—Distributing the Flow of Energy

The escapement acts like a brake system. It releases the stored energy from the gear train in small, precise increments instead of all at once. The most common type is the Swiss lever escapement. It allows the gear train to "tick" forward step by step.

Oscillator—Regulating Time

The oscillator is usually a balance wheel with a thin spiral spring (the hairspring). It swings back and forth at a steady frequency, much like a pendulum in a clock. Each swing controls the escapement, ensuring the gears move at a precise pace.

Time Display—The Hands

The controlled movement of the gears drives the watch hands: the seconds hand advances with each tick, while the minute and hour hands move more slowly through additional reductions in the gear train.

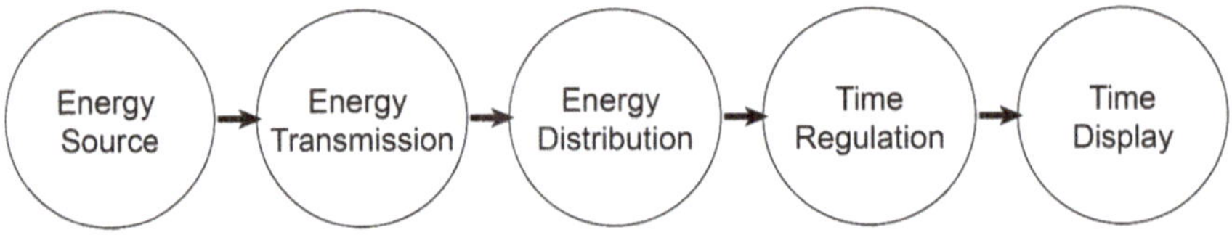

Fig. 1.2 Principle of a mechanical movement

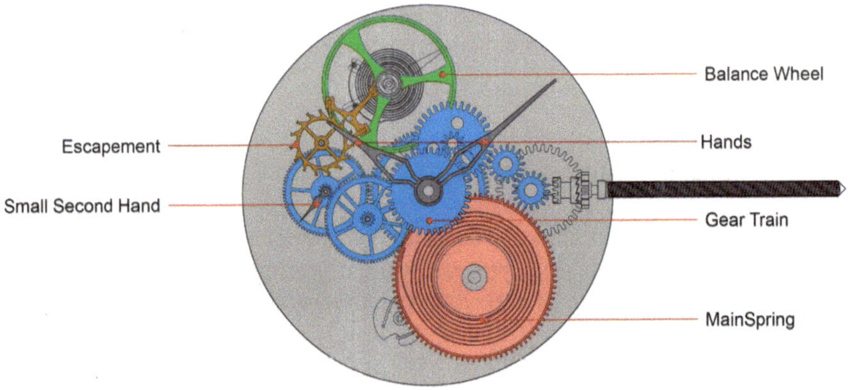

Fig. 1.3 Components of a hand-wound mechanical movement

Additional Mechanisms

Complications (extra features) such as date displays, chronographs, or moon phases are added through additional gear systems powered by the same mainspring.

A hand-wound mechanical watch, as depicted in Fig. 1.3, is powered by a mainspring that must be manually wound by turning the crown. As the spring unwinds, it powers the gears and hands. These watches are appreciated for their craftsmanship and traditional mechanics, but require regular winding, usually once a day [3].

Louis Audemars played an important role in the development of early crown-winding watch movements in the 19th century. In 1838, he created a mechanism that allowed a watch to be wound using the crown instead of a separate key. This innovation made watches more practical and convenient, since users no longer needed an external tool to wind or set the timepiece. The crown-winding system also improved reliability by reducing openings in the watch case through which dust and moisture could enter. Audemars' work marked a significant advancement in watchmaking and contributed to the modern winding systems still used in mechanical watches today [4].

An automatic mechanical watch, as presented in Fig. 1.4, also uses a mainspring, but it winds itself automatically using a rotor that moves with the wearer's wrist motion. This means no daily winding is necessary if worn regularly. However, if not worn for several days, it may stop and require manual winding or resetting.

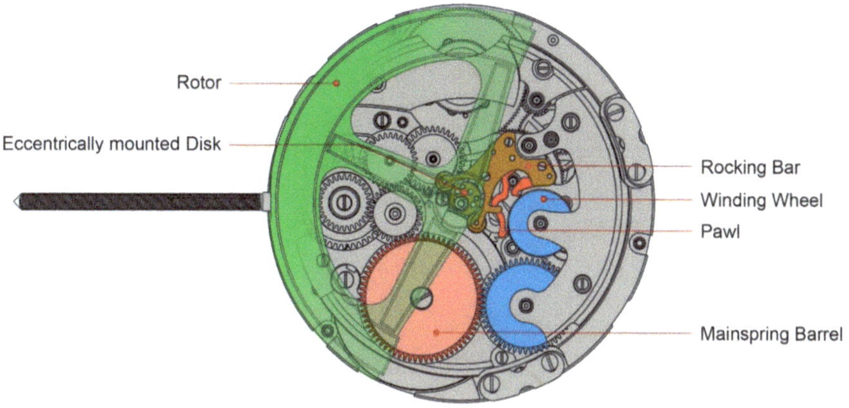

Fig. 1.4 Components of a mechanical automatic movement

Abraham-Louis Perrelet was a Swiss watchmaker from the 18th century who is widely credited with developing one of the first self-winding watch movements around 1770. His invention used the natural motion of the wearer's body to wind the mainspring automatically, eliminating the need for frequent manual winding. The mechanism worked through an oscillating weight that moved up and down while the wearer walked. Perrelet's innovation represented a major step forward in horology and laid the foundation for the automatic watches that became popular centuries later. His work greatly influenced later watchmakers and remains an important milestone in the history of mechanical watchmaking [5].

1.2.1.2 Tuning Fork Movements

A tuning fork movement is a type of electronic watch mechanism that was developed in the 1960s, most famously by Bulova in their "Accutron" watches. It represented a major step between purely mechanical watches and the later widespread quartz watches. By the 1970s, quartz watches (32,768 Hz crystals) became cheaper, more robust, and even more accurate, leading to the decline of tuning fork movements [6].

Tuning Fork

The heart of the movement is a small metal tuning fork, typically vibrating at 360 Hz (far higher than a traditional balance wheel's ~2.5–5 Hz).

Electromagnets and Coils

The tuning fork is kept vibrating by a pair of electromagnetic coils, powered by a small battery.

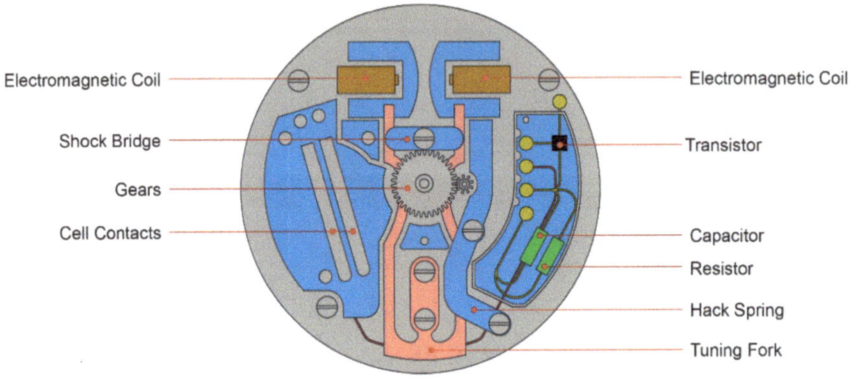

Fig. 1.5 Components of a tuning fork movement

Index Wheel

Attached to the tuning fork is a very fine mechanical system, including an index wheel with extremely small, delicate teeth.

Gears and Hand

The index wheel engages with the gear train that drives the watch hands, much like in a mechanical watch.

Unlike quartz watches, which "tick" once per second, a tuning fork watch's second hand moves in a smooth, continuous sweep, because the gear train is driven almost continuously by the vibrating fork.

Figure 1.5 provides an overview of the tuning fork movement elements.

1.2.1.3 Quartz Movements

The first quartz watch movement was developed by Seiko, specifically by its subsidiary Suwa Seikosha in Japan, in 1969. A quartz watch movement works by using the natural vibrations of a quartz crystal to keep very accurate time. Quartz vibrates at a highly stable and consistent frequency, which means quartz watches usually deviate by only a few seconds per month — much more precise than traditional mechanical watches [7].

Battery as a power source

A small battery supplies electrical energy to the watch. Unlike mechanical watches, no winding is necessary because the battery continuously powers the movement.

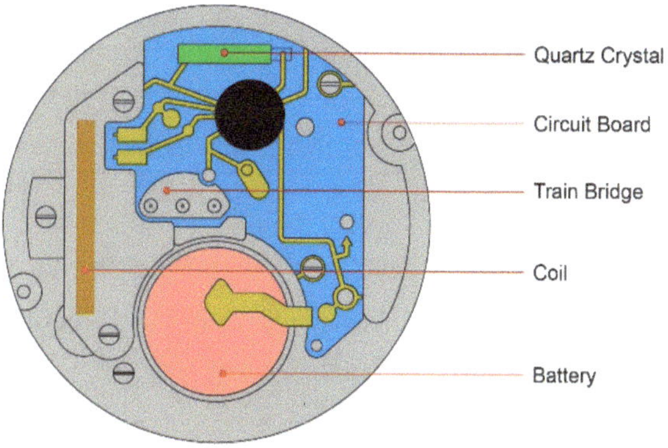

Fig. 1.6 Components of a quartz movement

Quartz crystal oscillator

The electricity from the battery passes through a tiny piece of quartz crystal, which is cut and shaped to vibrate at a precise frequency when voltage is applied. Quartz crystals typically vibrate at 32,768 times per second (32.768 kHz).

Frequency division

These very fast vibrations are too rapid to measure time directly. An electronic circuit reduces the oscillations down to a manageable rate—exactly one pulse per second.

Step motor

Each second, this electronic pulse is sent to a miniature stepping motor. The motor converts the electrical impulse into mechanical energy.

Gear train and display

The motor drives a system of gears that move the hands of the watch. These gears translate the motor's tiny steps into the continuous motion of the second, minute, and hour hands. Finally, the watch displays the time either on an analogue dial (with hands) or a digital display, depending on the design.

Figure 1.6 presents the components of a quartz movement.

1.2.1.4 Other Movements

Spring Drive

A Spring Drive movement is a unique watch technology developed by Seiko that combines the traditional mechanical mainspring with an advanced electronic regulation system.

Like a mechanical watch, a Spring Drive is powered by a mainspring, which stores energy when wound and gradually releases it through a gear train. However, instead of using a traditional escapement and balance wheel to regulate time, Spring Drive employs a device called the Tri-synchro Regulator.

The Tri-synchro Regulator controls energy in three forms:

- Mechanical energy—from the unwinding mainspring.
- Electrical energy—generated by a tiny rotor (glide wheel) spinning within a magnetic field.
- Electromagnetic energy—used to precisely regulate the glide wheel's speed.

The glide wheel rotates continuously and smoothly in one direction, rather than oscillating like a balance wheel. This motion generates a small electrical current, which powers a quartz crystal oscillator. The quartz provides an ultra-precise time reference, and an electromagnetic brake applies just the right amount of resistance to keep the glide wheel turning at exactly the correct speed.

Quartz/Solar

A solar-powered quartz watch has a rechargeable battery that is charged by light (sunlight or artificial). The watch works like a regular quartz movement but draws its energy from a solar panel, often hidden under the dial. These watches are eco-friendly and may run for months after a full charge.

Quartz/Radio Controlled

These radio-controlled quartz watches receive regular time signals via radio waves from atomic clocks, ensuring near-perfect accuracy. The quartz mechanism is still the timekeeper, but it's automatically corrected to the precise time based on the signal—often daily or several times a day.

Quartz/Kinetic

A kinetic watch combines mechanical motion and quartz precision. It generates electricity through a rotor that moves with the wearer's wrist, like an automatic watch. Instead of winding a spring, it charges a battery or capacitor that powers a quartz movement. This eliminates the need for battery replacements.

Smartwatch

A smartwatch is a digital or hybrid wearable device with advanced features beyond timekeeping. It typically includes functions like notifications, fitness tracking, GPS, heart rate monitoring, and even apps. Smartwatches are battery-powered (usually rechargeable) and connect to smartphones via Bluetooth or Wi-Fi.

1.2.2 Types of Watches by Watch Face

The watch face is the dial surface of a wristwatch that displays time and other functions, combining both technical readability and esthetic design [8].

Fig. 1.7 Traditional analogue watch face

Analogue/Traditional

A traditional analogue watch (see Fig. 1.7) has a classic dial with two or three hands (hour, minute, and sometimes second) moving over a face marked with numerals or indices. It's the most common and widely recognized watch style, offering an intuitive and elegant way to read the time.

Analogue/One Hand

A one-hand analogue watch (see Fig. 1.8) features only a single hand, usually for the hour. Time is read by estimating the position of the hand between hour markers, often subdivided into 15- or 5-minute increments. This minimalist design encourages a more relaxed perception of time and has a distinctive esthetic [9].

Fig. 1.8 Analogue one hand watch face

Analogue/Regulateur

A regulateur watch (see Fig. 1.9) separates the hour, minute, and second hands onto different sub-dials—typically, the minute hand remains central. In contrast, hours and seconds are shown on smaller, offset sub-dials. This design originated in observatories and master clocks for precision time reading, especially minutes.

Analogue/Rétrograde

A rétrograde display (see Fig. 1.10) shows time or other functions like date or day on a semi-circular scale. The hand moves along the arc and then snaps back to the beginning at the end of its cycle. This creates a dynamic and visually engaging dial layout often seen in high-end or artistic watches.

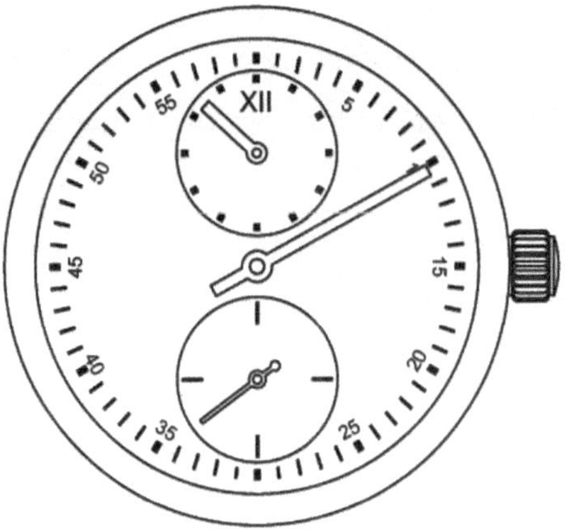

Fig. 1.9 Analogue regulateur watch face

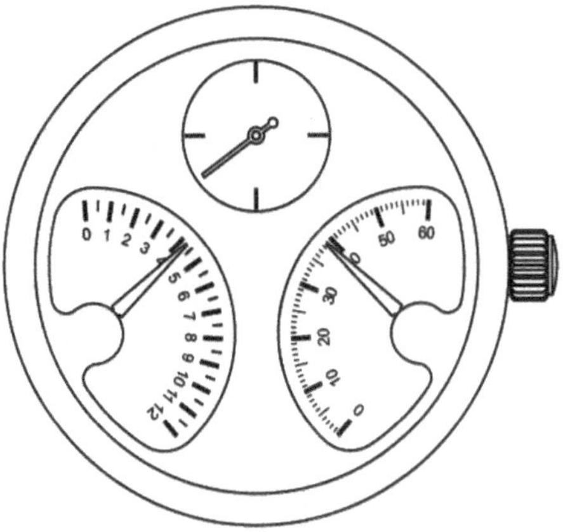

Fig. 1.10 Analogue rétrograde watch face

Fig. 1.11 Digital watch face

Digital

A digital watch (see Figure 1.11) displays time using numeric digits, typically on an LCD or LED screen, rather than using hands. This style often includes other features like a stopwatch, alarm, and backlighting, and can be more functional and sportier. Time is read directly, with no interpretation needed [10].

The concept of displaying time digitally—using numbers rather than traditional clock hands—dates back to the 1920s and 1930s, when mechanical "jump hour" watches were first developed. These watches used rotating discs to display hours and minutes through small windows, which could be seen as an early form of digital display, though purely mechanical.

The true digital watch emerged with the development of electronic components. In the 1960s, advances in transistor technology and the invention of the light-emitting diode (LED) enabled fully electronic time displays. In 1970, the Hamilton Watch Company introduced the first commercially available digital watch, the "Pulsar", which used an LED display that showed the time at the press of a button. This revolutionary design captured global attention, though the technology was expensive and battery life was limited. By the mid-1970s, liquid crystal display (LCD) technology began to replace LEDs, offering a continuous display without draining the battery as quickly. Companies like Casio, Seiko, and Timex popularized LCD digital watches, which became more affordable and widely available. This period also saw the introduction of additional functions like calculators, alarms, stopwatches, and even early programmable features.

Around 1973, Stowa developed a mechanical wristwatch with a digital display called the Stowa Digital Automatic. This watch was part of a broader trend in the early 1970s where mechanical watchmakers experimented with *"mechanical digital"* designs as a response to the rise of electronic digital watches.

1.2.3 Types of Watches by Complication

The term "complication refers to any feature in a watch that goes beyond the simple display of hours, minutes, and seconds. Complications demonstrate a brand's craftsmanship and prestige. Very expensive watches combine several complications; some of these valuable pieces are called "Grande Complication".

Date

Typically, through a small window (date aperture) on the dial, often at the 3 or 6 o'clock position, the date is displayed.

Mechanical wristwatches with a date display (see Fig. 1.12) are among the most common and practical expressions of traditional horology. While the date complication may appear simple to the wearer, it adds a meaningful layer of mechanical complexity to an otherwise straightforward movement.

Date + Day of the Week

Watches with this complication display the current date and day of the week, typically through two small windows at the 3 or 6 o'clock position (see Fig. 1.13).

Chronograph

A stopwatch function that allows the wearer to measure elapsed time using pushers that start, stop, and reset the central stop hand and additional sub dials. A traditional chronograph watch (see Fig. 1.14) features two separate pushers (buttons), usually located at 2 o'clock and 4 o'clock on the case. Some chronographs have quarter-minute scales, others have fifth-scale scales. This allows you to measure either a quarter or a fifth of a second.

The first true chronograph watch was developed in 1821. It's credited to Nicolas Mathieu Rieussec, a French watchmaker, who created a device to measure elapsed time during horse races. His invention used an inked marker that dropped a dot on

Fig. 1.12 Watch face with date complication

Fig. 1.13 Watch face with date + day of the week complication

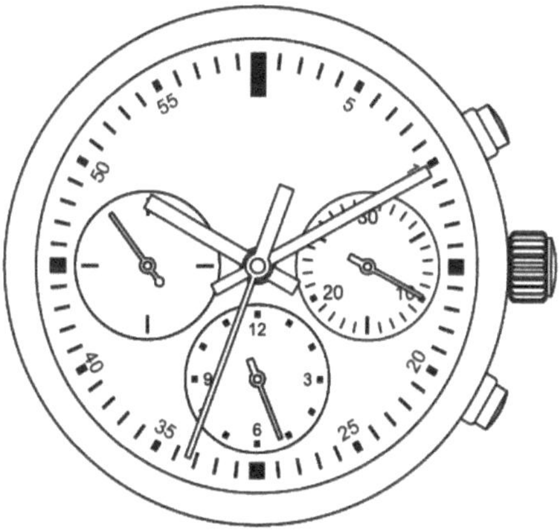

Fig. 1.14 Watch face with chronograph complication

a rotating dial to record timing—hence the name *chronograph*, from Greek *chronos* (time) and *graphein* (to write).

- Top pusher (2 o'clock): Starts and stops the chronograph.
- Bottom pusher (4 o'clock): Resets the chronograph to zero.

A flyback chronograph is a special type of chronograph that allows the user to reset and restart the timing function instantly with a single press of a button. In a conventional chronograph, the user must first stop the stopwatch, then reset it to zero, and finally start it again. A flyback mechanism combines these steps into one action.

When the flyback push-piece is pressed while the chronograph is running, the chronograph hands immediately return to zero and begin timing again without interruption. This function is especially useful for measuring consecutive intervals of time, such as navigation calculations in aviation or timing laps in motorsports.

The mechanism works through a sophisticated system of levers, cams, and hammers inside the movement. Activating the flyback function disengages the

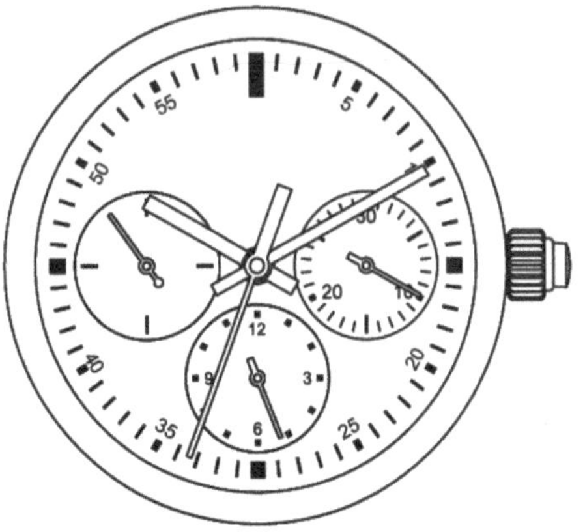

Fig. 1.15 Watch face with monopusher chronograph complication

chronograph, resets the hands, and reconnects the timing train almost instantly. Because of its technical complexity, the flyback chronograph is considered one of the most refined complications in mechanical watchmaking [11].

Monopusher Chronograph

A monopusher chronograph (see Fig. 1.15) uses only one pusher, typically located at 2 o'clock or integrated into the crown. The single pusher controls all three actions: start, stop, and reset. Usually, the sequence is:

- Press once to start.
- Press again to stop.
- Press a third time to reset.

Chronograph Rattrapante (Split-Seconds Chronograph)

An advanced chronograph with two central stop hands is called "Rattrapante". One can be stopped to measure intermediate times (split times), then catch up with the other hand when released.

A rattrapante chronograph (see Fig. 1.16), also known as a split-seconds chronograph, is one of the most sophisticated complications in mechanical wristwatches. It allows the measurement of two separate time intervals simultaneously using a single chronograph mechanism.

Minute Repeater

This highly complex complication chimes the time on demand using different tones for hours, quarter hours, and minutes, activated by a slide or button on the case.

Mechanical wristwatches with a minute repeater (see Fig. 1.17) are widely regarded as one of the highest achievements in traditional watchmaking. They combine mechanical complexity, acoustic engineering, and artisanal craftsmanship in a way that few other complications can match. Unlike clocks, wristwatches have extremely limited space, making sound production especially difficult [12].

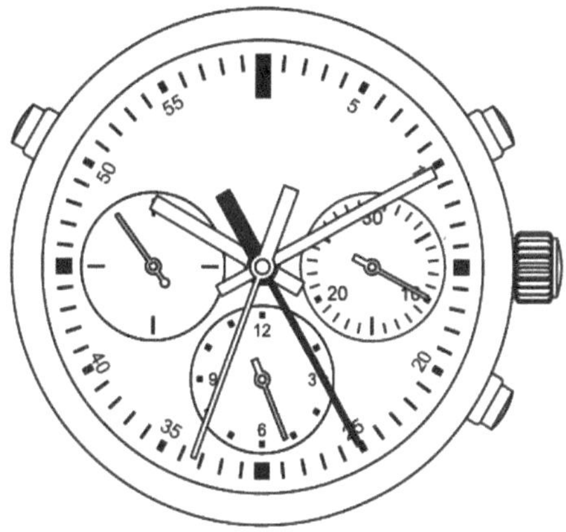

Fig. 1.16 Watch face with chronograph rattrapante complication

Fig. 1.17 Watch face with minute repeater complication

GMT/UTC

The GMT or UTC complication (see Fig. 1.18) allows tracking a second time zone, via an additional hand that points to a 24-hour scale or a 24-hour disk.

GMT stands for Greenwich Mean Time. It is a time standard originally based on the Earth's rotation and historically used as the world's reference time. GMT watches, commonly travel watches, are typically adjusted in whole-hour increments.

UTC is the abbreviation for Universal Time Coordinated, often referred to as "Zulu time," the modern global time standard. It is based on atomic clocks, not Earth's rotation. Pilots use UTC aviation watches as a common time reference to ensure safe and coordinated flight operations worldwide [13].

World Timer

A world timer displays the time in multiple time zones simultaneously, typically showing 24 cities representing each time zone on a rotating bezel or ring (see Fig. 1.19).

Fig. 1.18 Watch face with GMT/UTC complication

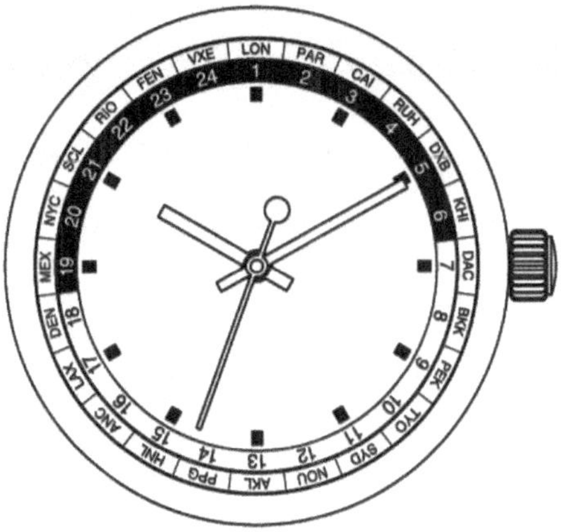

Fig. 1.19 Watch face with world timer complication. (The time zones are indicated by the IATA airport codes.)

The first wristwatch with a "world timer" display—meaning a mechanical watch that shows the time in multiple time zones simultaneously—is generally credited to Patek Philippe in the 1930s, based on a design by Swiss watchmaker Louis Cottier [14].

Moon Phase

This complication shows the current phase of the moon (e.g., new moon, full moon) in a sub dial or aperture, tracking the lunar cycle of roughly 29.5 days. The moon phase complication is one of the oldest and most emotional complications in watch-making because it combines mechanics, astronomy, and esthetics in a highly visible form. Moon phase displays on wristwatches range from simple decorative indicators to highly precise astronomical complications. The most common forms are the following [15]:

- Classical Moon Phase (See Fig. 1.20): A rotating disc beneath the dial shows two identical moons against a starry sky. As the disc turns, an aperture reveals different portions of the moon, simulating the lunar cycle from new moon to full moon and back again.

- Double Moon Phase (See Fig. 1.21): The so-called "Double Moon" display found in certain perpetual calendar watches from IWC Schaffhausen is a special form of moon phase indication that simultaneously depicts the lunar phase as seen from both hemispheres of Earth. A conventional moon phase display shows the moon as viewed from the northern hemisphere. However, the orientation of the moon appears visually inverted when observed from the southern hemisphere. IWC therefore developed a "Double Moon" display in which one moon is shown for the northern hemisphere and a mirrored moon is shown for the southern hemisphere.

Fig. 1.20 Watch face with classical moon phase complication

Fig. 1.21 Watch face with a double moon phase complication

- Pointer Moon Phase Moon Phase (See Fig. 1.22): Instead of a moon disc hidden behind an aperture, a hand points to the current lunar phase on a scale. The pointer moon phase indication is often used in modern independent designs.

Perpetual Calendar

A perpetual calendar (see Fig. 1.23) automatically accounts for the varying lengths of months and leap years, displaying the correct date, day, month, and often moon phase — in most cases without requiring adjustment until 2100.

The first known watch with a perpetual calendar is generally credited to Thomas Mudge, an English watchmaker, in the mid-18th century. The first perpetual calendar in a wristwatch was launched by Patek Philippe in 1941. The IWC perpetual calendar developed by Kurt Klaus (first introduced in 1985) stands out for relative user-friendliness compared with traditional perpetual calendars. The calendar mechanism is programmed to automatically recognise the varying lengths of months (30/31 days)

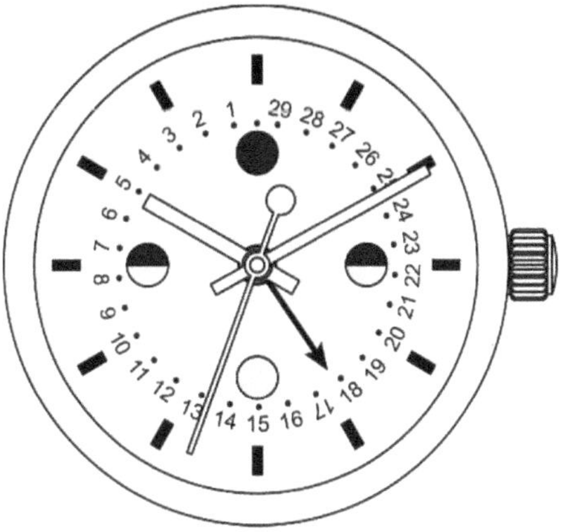

Fig. 1.22 Watch face with a pointer moon phase complication

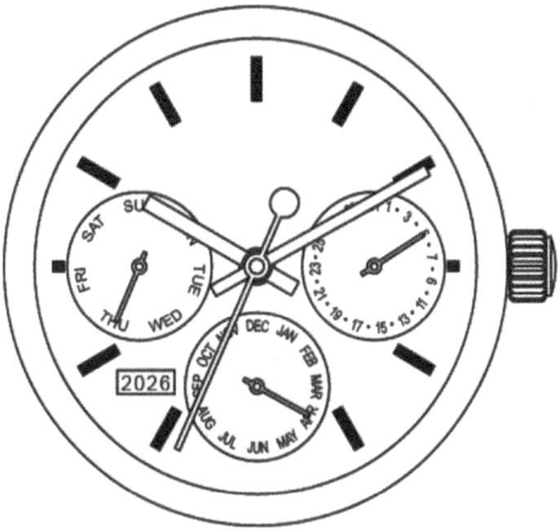

Fig. 1.23 Watch face with perpetual calendar complication

and leap years, inserting the extra day in February when required—all without manual correction until 2100 (when the Gregorian calendar omits a leap year) [16].

Annual Calendar

An annual calendar is similar to a perpetual calendar, but it requires manual adjustment once per year—at the end of February. It accounts for 30- and 31-day months.

Compared to a simple date watch, the annual calendar (see Fig. 1.24) requires correction only once per year, saving time and reducing the risk of misadjustment. The calendar display is often integrated into the dial in sub-dials or apertures, sometimes combined with moon phases or day-of-week indicators.

Power Reserve

A power reserve display indicates how much longer the watch will run before needing to be wound or recharged, usually shown through a scale or gauge on the dial.

A power reserve indicator (see Fig. 1.25) transforms a purely mechanical watch into a more user-friendly instrument, providing real-time insight into the mainspring's energy. It enhances practicality, prevents mishandling, and adds an elegant visual element, making it both a functional and esthetic feature of modern mechanical watchmaking [17].

Jumping Hour

Jumping hour stands for a digital-style complication where the hour is displayed through a window and "jumps" to the next number instantly at the top of each hour, rather than moving gradually.

Mechanical wristwatches with a jumping hour complication (see Fig. 1.26) represent a distinctive and intellectually elegant approach to displaying time. Jumping hours appeared as early as the late 19th and early 20th centuries, particularly in

Fig. 1.24 Watch face with annual calendar complication

Fig. 1.25 Watch face with power reserve complication

pocket watches and early wristwatches. They were admired for their legibility and novelty but were technically complex and costly to produce.

Tourbillon

A tourbillon is a rotating cage mechanism that houses the escapement and balance wheel, designed to counteract the effects of gravity on accuracy. It's often visible through an opening on the dial.

The tourbillon was invented in 1801 by Abraham-Louis Breguet (founder of Breguet)—albeit for pocket watches.

In wristwatches, the tourbillon (see Fig. 1.27) is often placed prominently on the dial side, allowing the wearer to observe its continuous motion. This visual element has become a hallmark of *haute horlogerie* [18].

The following types of tourbillons can be distinguished [19]:

- Classic Tourbillon

Fig. 1.26 Watch face with jumping hour complication

Fig. 1.27 Watch face with tourbillon complication

The classic (or traditional) tourbillon consists of a rotating cage that carries the escapement, balance wheel, and hairspring. The cage is typically supported by bridges on both the top and bottom and usually completes one full rotation per minute.

- Flying Tourbillon

  The flying tourbillon was invented by Alfred Helwig in 1920. Unlike the classic design, the cage is supported from only one side (usually underneath), eliminating the upper bridge. This creates the visual impression that the mechanism is "floating" in space.

- Multi-Tourbillon (Double, Triple, Quadruple Tourbillon)

  A multi-tourbillon watch contains more than one tourbillon mechanism. Examples include double, triple, or even quadruple tourbillons. The tourbillons may be connected through a differential system that averages their rates.

- Multi-Axis Tourbillon

  A multi-axis tourbillon (sometimes called a gyro-tourbillon) uses a single regulating organ mounted within multiple rotating cages that rotate on different axes. Instead of having several separate tourbillons, one tourbillon rotates simultaneously in two or three dimensions. These Tourbillons are designed to compensate for gravitational effects from many more positions than a conventional tourbillon.

Tide Indicator

A tide indicator displays the state of the tide (high/low) at a specific location, useful for sailors and fishermen.

Tide indicator watches (see Fig. 1.28) are defined by their close connection to nature, precision, and maritime functionality. They are designed to display tidal

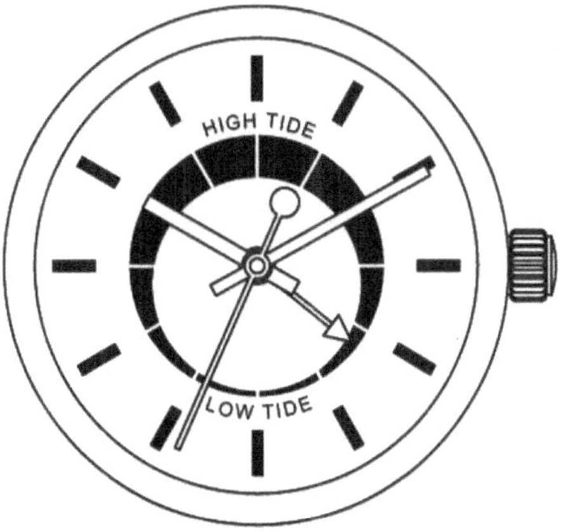

Fig. 1.28 Watch face with tide indicator complication

movements, making them valuable companions for sailors, surfers, fishermen, and coastal explorers who rely on the rhythm of the sea. Their character combines technical accuracy with practical design, often featuring clear displays and robust construction suited for marine environments. Tide indicator watches reflect a spirit of freedom, adventure, and respect for natural cycles, blending utility with a distinctive nautical esthetic.

Alarm

Alarm watches (see Fig. 1.29) sport the mechanical or quartz function that sets and sounds an audible alarm at a chosen time.

The Vulcain "Cricket" is one of the historically most famous alarm watches, known for its spring-powered alarm that "*chirps*" like its namesake insect. Originally released in the 1940s, it became popular among U.S. Presidents and collectors [20].

Fig. 1.29 Watch face with alarm complication

Fig. 1.30 Watch face with siderial time complication

Sidereal Time

As the name suggests, this complication displays sidereal time, which is a method of timekeeping based on the Earth's rotation relative to the fixed stars, rather than the Sun.

The first wristwatch with a sidereal time display was developed by Patek Philippe in 2001. The display of the "Sky Moon Tourbillon" was provided by a rotating starry sky on the back, which showed the celestial movement over a chosen location.

The "Portugieser Sidérale Scafusia" by IWC displays solar time (ordinary civil time) and sidereal time, which is about 4 minutes shorter than a solar day and is used by astronomers to track the position of stars.

A 24-hour day (more precisely called the solar day) and a sidereal day are both ways of measuring Earth's rotation, but they use different reference points. The sidereal day is shorter because Earth moves along its orbit around the Sun while rotating on its axis.

The case back reveals a customized celestial chart showing the night sky from a specific location chosen by the owner. It also displays sunrise and sunset times and other astronomical data. The watch was developed mainly by Jean-François Mojon [21].

Figure 1.30 demonstrates the typical layout of a watch displaying sidereal time next to local time.

1.2.4 Types of Watches by Style

Style categories are one of the main ways enthusiasts, collectors, and brands organize watches. While there's some overlap, most watches fall into one or more recognized style groups based on their design, function, and intended use [22].

Casual Watches

Characteristics:

- Designed for everyday wear
- Simple, functional design
- Comfortable materials (e.g., leather, fabric, rubber)
- Easy-to-read dials

Casual watches (see Fig. 1.31) are defined by their relaxed, versatile, and approachable character. They are designed to fit naturally into everyday life, bridging the gap between purely functional timepieces and more formal or luxury-oriented watches. A casual watch does not demand attention or strict dress codes; instead, it complements a wide range of outfits, from jeans and T-shirts to business-casual attire. In terms of design, casual watches typically emphasize clarity, comfort, and practicality. Dials are easy to read, case sizes are moderate, and materials are chosen for durability rather than opulence. Stainless steel, fabric, or leather straps, along with simple color palettes, are common, giving these watches a timeless and unpretentious appearance. The overall character of a casual watch is one of effortlessness and reliability. It is a watch designed for daily wear, without concern for special occasions or delicate handling. Casual watches often reflect a lifestyle that values authenticity and functionality over status symbols, making them personal companions rather than statements of prestige. In this sense, they are less about showcasing luxury and more about expressing ease, individuality, and everyday confidence.

Dress Watches

Characteristics:

- Elegant, minimalist design
- Slim profile (fits under shirt cuffs)
- Leather straps are common
- Often feature a clean dial with few complications
- Focus on sophistication over function

Dress watches (see Fig. 1.32) are defined by their refined, elegant, and understated character. They are designed to complement formal attire rather than dominate it, serving as a subtle expression of taste and sophistication. A true dress watch is discreet and harmonious, blending seamlessly with suits, tuxedos, or other formal clothing without drawing unnecessary attention.

Concerning design, dress watches prioritize simplicity and proportion. Slim cases, clean dials, and restrained details are essential elements, allowing the watch to slip easily under a shirt cuff. Precious metals or finely finished stainless steel are commonly used, paired with high-quality leather straps. Complications are kept to a minimum—often limited to hours, minutes, and seconds, or a discreet date—because visual clarity and balance take precedence over technical display.

Fig. 1.31 Casual watch

The character of a dress watch is one of timeless elegance and quiet confidence. It reflects an appreciation for tradition, craftsmanship, and restraint rather than innovation or ruggedness. Unlike more casual or sporty timepieces, a dress watch is not designed for versatility or durability in demanding environments; instead, it excels in moments where refinement and good manners matter most. As such, it becomes a symbol of classic style, respect for formality, and enduring esthetic values.

Fashion watches

Characteristics:

- Trend-focused, style-driven
- Designed by fashion brands rather than watchmakers
- Emphasis on appearance rather than technical quality
- May include bold colors, large dials, or decorative elements
- Often mass-produced and affordably priced

Fashion watches (see Fig. 1.33) are defined by their expressive, trend-driven, and visually oriented character. Their primary purpose is not long-term horological value or technical innovation, but style and visual impact. They function as accessories first

Fig. 1.32 Dress watch

and timekeeping instruments second, often designed to reflect current fashion trends, seasonal colors, or a specific lifestyle image.

Fashion watches tend to be bold, playful, or deliberately minimalist, depending on prevailing trends. Unconventional case shapes, striking colors, oversized proportions, or graphic dials are common, as are materials chosen for their aesthetic appeal rather than durability or traditional watchmaking prestige. Branding often plays a central role, with logos and recognizable design signatures serving as key elements of the watch's identity.

The overall character of a fashion watch is dynamic and short-lived, closely tied to changing tastes and cultural moments. These watches are meant to make a statement, complete an outfit, or express personal style rather than demonstrate craftsmanship or heritage. As a result, fashion watches emphasize accessibility and visual relevance, appealing to wearers who see a watch as a reflection of mood, trend awareness, and individual expression rather than a timeless object to be worn for decades.

Luxury Watches

Characteristics:

- High-end craftsmanship and materials (gold, sapphire crystal, etc.)
- Produced by renowned watchmakers
- Often include mechanical movements

Fig. 1.33 Fashion watch

- Expensive and often seen as status symbols
- Attention to detail and heritage

Luxury watches (see Fig. 1.34) embody a distinctive character that combines precision, heritage, and refined elegance. They represent far more than instruments for measuring time; they are symbols of craftsmanship, status, and personal expression. Each luxury watch reflects meticulous attention to detail, from the quality of the materials to the complexity of the movement within. Tradition and innovation exist side by side, as centuries-old watchmaking techniques are often paired with modern technology and contemporary design. The character of luxury watches is defined by exclusivity and longevity, created to endure both in function and esthetic appeal. They convey confidence, taste, and an appreciation for artistry, appealing to individuals who value excellence, authenticity, and timeless sophistication.

Sports Watches

Characteristics:

Fig. 1.34 Luxury watch

- Built for durability and performance
- Features like water resistance, shock resistance, and chronograph functions
- Made with tough materials (stainless steel, titanium, ceramics)
- Often larger and more rugged
- May include diving, pilot, or racing styles

Sports watches (see Fig. 1.35) are characterized by strength, functionality, and a spirit of adventure. They are designed to perform reliably under demanding conditions, combining durability with precision and clarity. Their character is shaped by robust materials, practical features such as water resistance, shock protection, and high legibility, and a design that prioritizes performance without sacrificing style. Sports watches embody an active lifestyle and a sense of readiness, appealing to individuals who value resilience, versatility, and confidence. They are not only tools for measuring time but dependable companions that reflect energy, determination, and a passion for movement and challenge.

Skeleton Watches

Characteristics:

- Exposed inner mechanics; parts of the movement are visible
- Designed to showcase watchmaking craftsmanship

Fig. 1.35 Sports watch

- Artistic, intricate, and often luxurious
- Can be mechanical or automatic
- Often considered conversation pieces

Skeleton watches (see Fig. 1.36) are defined by a character of transparency, artistry, and mechanical fascination. By revealing the inner workings of the movement, they celebrate the beauty of engineering and the complexity of timekeeping. Their character is shaped by precision craftsmanship, where every visible component is carefully finished and intentionally displayed. Skeleton watches balance technical mastery with bold design, creating a sense of depth and visual intrigue. They appeal to individuals who appreciate innovation, detail, and the harmony between form and function, expressing a confident, avant-garde style and a deep admiration for the art of watchmaking.

Fig. 1.36 Skeleton watch

1.2.5 Types of Watches by Specific Purpose

The classification of watches by their special purpose is based on the environments, professions, or activities for which they were designed. Many of these special-purpose watches were originally tool watches: practical instruments developed to serve a function beyond just telling the time [23].

Divers' Watches

- Purpose: Underwater use for divers.
- Features: High water resistance (200m+), unidirectional bezel for timing dives, luminous hands/indices, robust cases.

Divers' watches (see Fig. 1.37) are specialized timepieces designed to withstand the extreme conditions encountered underwater, combining robust engineering with practical functionality. One of their defining characteristics is water resistance, typically rated in meters or atmospheres, which ensures that the watch remains operational even at significant depths. To achieve this, dive watches often feature screw-down crowns, reinforced gaskets, and corrosion-resistant materials such as stainless steel, titanium, or ceramics. Another hallmark is the unidirectional rotating

Fig. 1.37 Diver's watch

bezel, which allows divers to track elapsed time safely. Because it only turns in one direction, any accidental movement can only shorten, rather than extend, the measured dive time, providing a critical safety feature. The dial and hands are usually equipped with highly luminescent markers, ensuring excellent readability in low-light or murky underwater conditions. Durability extends beyond water resistance, with dive watches engineered to resist shocks and pressure changes. The crystals are typically made from scratch-resistant sapphire or hardened mineral glass, capable of enduring harsh underwater environments. Many dive watches also include a helium escape valve, which is essential for professional saturation divers to prevent damage during decompression. Beyond functionality, divers' watches often have a robust and secure strap or bracelet, sometimes made of rubber or metal, that can fit over a wetsuit [24].

ISO 6425 is the international standard that defines the requirements for divers' watches. Its purpose is to ensure that a watch marketed as a "diver's watch" is suitable for underwater use and meets strict safety and performance criteria. ISO 6425 specifies the requirements and test methods for diving timepieces designed to withstand underwater use, including pressure, temperature changes, corrosion, shocks, and low visibility.

Key Points of ISO 6425:

- Water resistance: A diver's watch must withstand water pressure corresponding to at least 100 meters depth, and it must be tested under higher pressure than the stated rating.
- Reliability under harsh conditions: The watch must remain functional after exposure to temperature changes, shocks, and magnetic fields. Materials and construction must withstand saltwater exposure without significant corrosion or malfunction.
- Time measurement underwater: It must include a unidirectional rotating bezel or another device to safely measure elapsed dive time, preventing accidental extension of dive duration. The bezel must have clear markings, including a minute scale and a prominent zero marker.
- Legibility: The dial must be clearly readable in darkness and underwater, with luminous markers visible from a specific distance. A visible running indicator (often the seconds hand) has to show that the watch is operating.
- Resistance to condensation: The watch must not develop internal fogging when exposed to sudden temperature shifts.
- Secure construction: The strap or bracelet must be strong enough to prevent loss during a dive.
- Marking: Watches meeting the standard may be labelled as "Diver's Watch" followed by the depth rating (e.g., "Diver's 200 m").

ISO 6425 ensures that certified divers' watches are not just water-resistant but truly designed for professional and safe underwater diving conditions. Watches are generally considered "waterproof" if they can withstand a water depth of 1 meter for at least 30 minutes (DIN 8310) without damage for the entire lifetime of the product. Because this standard is practically impossible to guarantee, the term "waterproof" is generally prohibited or discouraged in many countries and industry standards.

Pilots' Watches

- Purpose: Pilot timekeeping.
- Features: Large, highly legible dials, oversized crown (usable with gloves), GMT or chronograph functions, anti-magnetic protection.

Pilots' watches are timepieces designed specifically for aviators, emphasizing legibility, reliability, and practical functionality in the demanding environment of flight. One of their most distinctive features is the large, highly readable dial, often with bold Arabic numerals and luminous hands and markers, ensuring that the time can be read quickly and accurately, even in low-light conditions or turbulent situations. Many pilots' watches also include a triangle at the 12 o'clock position, a traditional design element that allows pilots to orient the watch at a glance.

Accuracy and reliability are central to pilots' watches. These watches often feature precision automatic or manual movements capable of withstanding changes in altitude, temperature, and atmospheric pressure. Anti-magnetic cases are sometimes incorporated to prevent interference from cockpit instruments.

Functionality is further enhanced with features such as chronographs, which allow pilots to measure elapsed time, and UTC displays, enabling tracking of multiple time zones—a critical tool for long-haul flights. The crowns are often oversized, designed to be operable while wearing gloves, reflecting the practical needs of aviators.

Durable materials such as stainless steel, titanium, or scratch-resistant sapphire crystal ensure that pilot watches can endure the rigors of aviation. Straps are typically made of leather or fabric for comfort and secure wear during long flights. Beyond their technical advantages, pilots' watches carry a strong esthetic heritage, evoking the golden age of aviation and the adventurous spirit of early pilots, making them both functional instruments and iconic symbols of style.

The most well-known pilot's watches are probably the so-called B-Uhr watches. The B-Uhren (short for the German word *Beobachtungsuhren*, meaning "observation watches") were a special type of military timepiece developed in Germany in the late 1930s and used primarily during the Second World War.

B-Uhren were created for the Luftwaffe, the German air force. Their main purpose was to provide a precise and easily readable time for navigators aboard long-range military aircraft. Accurate timekeeping was essential for aerial navigation, especially when using celestial navigation and dead-reckoning methods over long distances and open sea.

The design of the B-Uhr was strictly standardized by the German military. Key features included:

- A large case, typically around 55 mm in diameter, to ensure excellent legibility
- A black dial with stark white markings
- Luminous hands and numerals for use in low-light conditions
- A prominent triangle with two dots at 12 o'clock to allow quick orientation
- An extra-long leather strap, enabling the watch to be worn over a flight jacket
- A hacking movement, allowing the seconds hand to be stopped for precise Synchronization

Two main dial types existed:

- **Type A** (earlier): hours marked 1–11 with the triangle at 12
- **Type B** (later): an outer minute scale (5–55) with a smaller inner hour circle

Only five companies were officially authorized to produce B-Uhren:

- Lange & Söhne
- Laco (Lacher & Co.)
- Stowa (Walter Storz)
- Wempe (using movements by Thommen)
- IWC (International Watch Company, Schaffhausen)

Each watch was individually numbered and engraved on the case back with military designations.

B-Uhren were not issued to ordinary pilots but to navigators, who often used them in conjunction with chronometers and navigation tables. They were considered

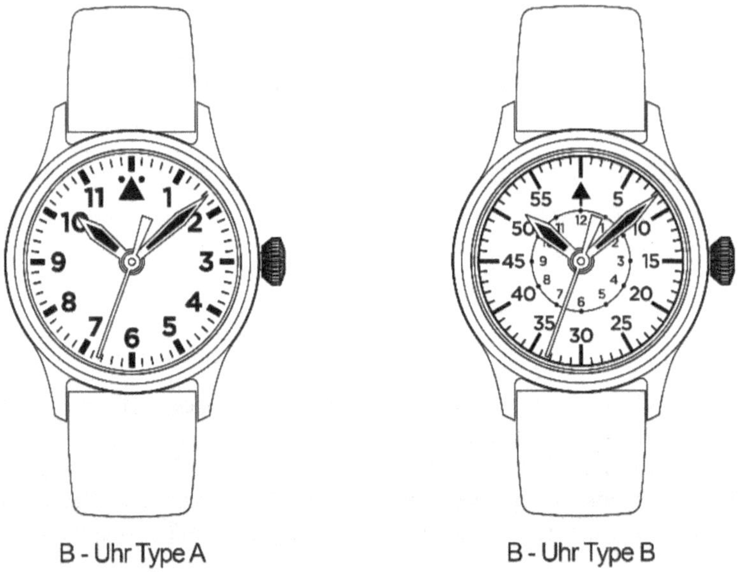

Fig. 1.38 Pilot's watches in B-Uhr style

precision instruments, technically owned by the state and subject to strict handling rules. In many cases,

After the war, many original B-Uhren were lost, destroyed, or dismantled. Surviving examples became highly sought-after collectors' items. In the late 20th and early 21st centuries, the distinctive design of the B-Uhr inspired numerous modern pilot watches, produced both by the original manufacturers and by many other brands [25].

Today, B-Uhren (see Fig. 1.38) are remembered not only as historical military instruments but also as icons of functional watch design, emphasizing clarity, precision, and purpose over decoration.

Railroad Watches

- Purpose: Ensure precise timekeeping for train conductors and engineers.
- Features: High accuracy, bold numerals for quick legibility, resistant to magnetism, regulated movements.

Railroad watches (see Fig. 1.39) are specialized timepieces developed in the 19th and early 20th centuries to ensure precision and safety in railway operations. During this era, accurate timekeeping was critical to prevent collisions and maintain reliable train schedules, making these watches essential tools for conductors, engineers, and other railway personnel. One of the most defining features of railroad watches is their highly accurate mechanical movement, often certified as a chronometer, which could maintain consistent time over long periods and under varying conditions.

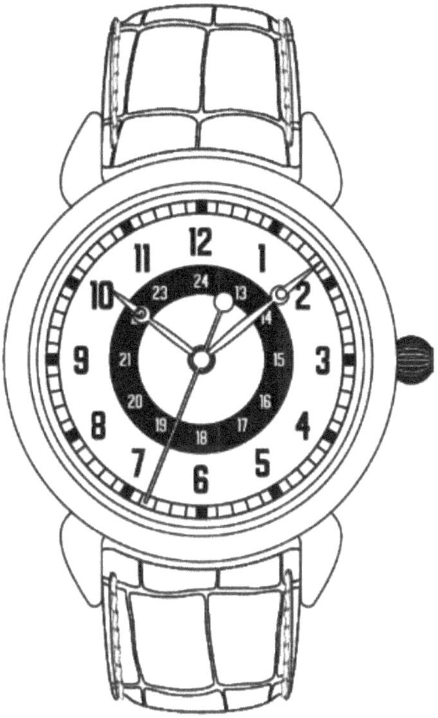

Fig. 1.39 Railroad watch

Legibility was another critical requirement. Railroad watches typically feature large, bold Arabic numerals, clear minute markers, and contrasting hands, allowing the wearer to read the time instantly, even in dim light or under stress. Many also employed open-face designs with the crown at 12 o'clock for quick and intuitive reading, as well as railroad-standard dial layouts that eliminated ambiguity, such as using full minute tracks instead of partial markers.

Durability and reliability were equally important. Railroad watches were built with robust cases, often in gold or nickel, and high-quality jewels in the movement to reduce wear and maintain accuracy over time. Some models included lever-set mechanisms, a safety feature that prevented accidental changes to the time, which could otherwise disrupt train schedules.

Overall, railroad watches combine precision, legibility, and robust engineering, reflecting a time when the safety of thousands of passengers depended on the dependable performance of a single timepiece. Today, they remain iconic symbols of horological craftsmanship and a reminder of the critical role of accurate timekeeping in the history of transportation.

Racing/Motorsport Watches

- Purpose: Timing motorsport events and speed measurement.
- Features: Chronograph function, tachymeter scale (to calculate speed), bold sporty design.

Racing or motorsport watches (see Fig. 1.40) are designed specifically for drivers, pit crews, and motorsport enthusiasts, combining precision, durability, and highly legible displays to measure short intervals of time with accuracy. One of their defining features is the chronograph function, which allows the wearer to time laps, calculate speeds, and track split-second intervals—a critical tool in racing environments. These watches often feature tachymeter scales on the bezel or dial, enabling drivers to calculate average speed over a known distance at a glance.

Legibility under fast-paced conditions is another hallmark. Racing watches typically have bold, contrasting numerals and hands, sometimes accented with bright colors, ensuring the time and chronograph measurements can be read instantly, even

Fig. 1.40 Racing watch

while moving at high speeds. The dials are often inspired by automotive instruments, with sub dials arranged for intuitive reading of elapsed minutes, seconds, and sometimes tenths of a second.

Durability and performance are emphasized through robust cases made from stainless steel, titanium, or lightweight alloys, as well as scratch-resistant sapphire crystals that can withstand vibrations, shocks, and temperature changes encountered on the track. Some models include pushers that can be operated while wearing racing gloves, reflecting the practical needs of professional drivers.

Beyond technical functionality, motorsport watches convey a sense of speed, precision, and style, often drawing esthetic cues from racing dashboards, steering wheels, and sports cars. They are not only reliable instruments for timing and calculation but also symbolic accessories that celebrate the thrill and heritage of automotive racing.

Medical Watches

- Purpose: Assist doctors and nurses.
- Features: Pulsometer scale (to measure heart rate), highly legible dials, sometimes designed to be worn upside down (for quick reading on patients).

Medical or doctors' watches (see Fig. 1.41) are specialized timepieces designed to assist healthcare professionals in performing precise medical tasks, particularly in measuring patients' vital signs. One of their most distinctive features is the pulsometer scale, which allows doctors and nurses to calculate a patient's heart rate quickly and accurately. By counting the number of beats over a specific interval, often 15 or 30 s, and reading the corresponding value on the scale, medical personnel can determine the pulse without additional equipment.

Legibility is paramount in a medical watch. These watches typically feature clear, large numerals and contrasting hands, ensuring that time can be read at a glance, even in stressful or low-light environments such as hospitals or emergency rooms. Many models also have a highly precise second hand, essential for timing heartbeats or respiratory rates accurately.

Durability and hygiene are also key considerations. Medical watches are often built with stainless steel or other corrosion-resistant materials, and some feature easily cleanable straps, reflecting the need for frequent sanitation in clinical settings. Certain designs are worn on the wrist upside down or with a flip design, allowing the wearer to read the dial quickly when performing examinations or taking vitals.

Overall, doctors' watches combine precision, clarity, and practicality, functioning as indispensable tools for healthcare professionals. They represent a unique intersection of horology and medicine, where careful timekeeping directly contributes to patient care and safety.

Sailing/Regatta Watches

- Purpose: Yacht racing and nautical navigation.
- Features: Regatta countdown timers, tide indicators, water resistance, robust construction.

Fig. 1.41 Medical watch

Sailing or regatta watches (see Fig. 1.42) are specialized timepieces designed for the unique demands of competitive sailing, combining precise timing functions with durability and readability in challenging marine environments. One of their most distinctive features is the regatta countdown timer, which allows sailors to track the crucial interval before a race starts. This function is often paired with audible or visual signals, ensuring that the crew can synchronize their manoeuvres precisely at the starting line.

Legibility is essential on a sailing watch. These watches typically feature large, clear numerals, high-contrast hands, and luminous markers, making it easy to read the time quickly, even in bright sunlight or low-light conditions at sea. Many models also include additional functions such as chronographs, dual time zones, or tide indicators, which help sailors plan and adjust their strategies according to local conditions.

Durability and water resistance are central to the design of sailing watches. They are commonly built with robust cases in stainless steel, titanium, or lightweight

Fig. 1.42 Regatta Watch

composites, coupled with scratch-resistant sapphire crystals and high water-resistance ratings to withstand splashes, immersion, and the corrosive effects of saltwater. Straps are often made from rubber or textile for comfort, flexibility, and secure wear during vigorous activity on deck.

Beyond practical functionality, sailing watches often reflect a nautical esthetic, with design cues inspired by marine instruments, ropes, and the colours of the sea. They combine precision timing, resilience, and readability, making them essential tools for competitive sailors while also embodying the spirit and elegance of life on the water.

1.2.6 Types of Watches by Target Groups' Gender

Watches designed specifically for women and specifically for men can be differentiated according to the following criteria [26]:

Case Size and Diameter

- Men's watches usually have larger cases, often ranging from 40 mm to 46 mm or more.
- Women's watches are typically smaller, often between 26 mm and 36 mm.

Strap/Band Width

- Men's watches have broader, more robust straps to match the larger case.
- Women's watches tend to feature narrower, more delicate bands.

Design and Style

- Men's watches often emphasize functionality, with bold or minimalist designs.
- Women's watches frequently incorporate more decorative elements—such as gemstones, mother-of-pearl dials, or elegant shapes.

Weight and Thickness

- Men's watches are usually heavier and thicker, designed to feel substantial.
- Women's watches are lighter and slimmer for comfort and esthetics.

Color Schemes

- Men's watches often use neutral, classic tones like black, silver, gold, navy, or grey.
- Women's watches may feature softer palettes (rose gold, white, pastel tones) or jewelry-like finishes.

Functionality and Complications

- Men's watches often highlight chronographs, tachymeters, diving bezels, or GMT functions.
- Women's watches usually keep the complications simpler, focusing more on style—though this has been changing in recent years, as many women now prefer technical timepieces.

The requirements men and women have for wristwatches often differ due to a combination of anatomical, stylistic, and lifestyle factors, although individual preferences increasingly overlap. From a physical perspective, men generally tend to prefer larger and heavier watches, as these better match broader wrists and convey a sense of robustness. Women, on average, often look for slimmer and lighter watches that sit comfortably on smaller wrists and feel less intrusive during daily wear. This leads to differences in case size, thickness, and overall proportions.

Esthetic expectations also play an important role. Men's watches are frequently designed with a focus on functionality, technical complexity, and a more restrained or utilitarian appearance. Features such as chronographs, rotating bezels, or high water-resistance are often valued, as they suggest performance and durability. Women's watches, by contrast, are more often regarded as both timekeeping instruments and pieces of jewellery. As a result, design elements such as elegant shapes, refined dials, decorative details, or the use of precious materials and gemstones tend to be more prominent.

Another difference lies in how the watch is integrated into personal style and occasions. Men commonly seek versatility, wanting a single watch that can transition easily from professional to casual settings. Women are more likely to view watches as accessories that complement different outfits, leading to a greater emphasis on color options, interchangeable straps, and stylistic variety. Despite these traditional distinctions, modern watch design increasingly reflects a shift toward unisex models, recognising that many women appreciate technical features and larger cases, while many men value minimalism, elegance, and slimmer profiles.

Recently, many unisex designs have emerged, blurring these distinctions. Some women prefer oversized men's watches, while some men opt for slimmer, more elegant pieces. The differences between watches designed for men and women are based primarily on different wrist diameters. The diagram below (Fig. 1.43) is based on anthropometric surveys. It shows a comparison of the wrists of men and women with heavy, average, and petite builds.

1.2.7 Types of Watches by Design Era

The design of wristwatches has changed repeatedly over time. The typical characteristics of each style era are reflected in the shape of the case and the dial graphics. Some stylistic features have returned to fashion over time.

Art Nouveau/Jugendstil (ca. 1890–1910)

Typical Elements of an Art Nouveau Watch (see Fig. 1.44):

- Flowing, organic lines.
- Stylized floral, plant, and feminine motifs.
- Emphasis on elegance and fluidity.
- Integration of watch and bracelet into a single, decorative whole.

Art Nouveau, known as Jugendstil in German-speaking countries, was an influential design movement that flourished between the late 19th and early 20th centuries. Characterized by organic, flowing lines, natural forms, and intricate ornamentation, it sought to unify art, design, and everyday life. Inspired by plants, flowers, and the curves of the natural world, Jugendstil appeared across architecture, furniture, jewelry, glass, and graphic arts. Its emphasis on craftsmanship, elegance, and harmonious design challenged industrial mass production, creating visually striking objects that combined beauty with functionality. The movement left a lasting impact on modern design, bridging traditional artistry with the esthetics of the modern era [27].

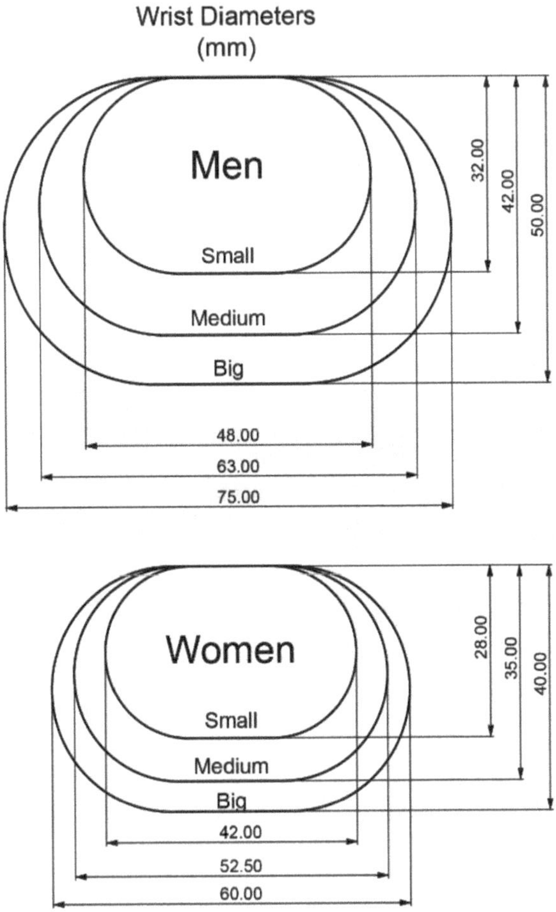

Fig. 1.43 Cross-sections of male and female wrists with measurements

Art Deco (ca. 1920–1940)

Typical Elements of an Art Deco watch (see Fig. 1.45):

- Geometric, streamlined shapes.
- Bold contrasts, symmetry, and repetition.
- Decorative yet modern, glamorous, and urban esthetic.
- Watches are often rectangular (tank style) or with decorative, jewel-like cases.

Art Deco was a prominent design movement that emerged in the 1920s and 1930s, characterized by geometric shapes, bold lines, and lavish ornamentation. It combined modernity with elegance, incorporating materials such as chrome, glass,

Fig. 1.44 Wristwatch in Art Nouveau style

and exotic woods to create a sense of luxury and sophistication. Influencing architecture, fashion, jewelry, and industrial design, Art Deco celebrated symmetry, streamlined forms, and decorative patterns, leaving a lasting mark on the visual language of the early 20th century [28].

Bauhaus (1919 to today)

Typical Elements of a Bauhaus-Style watch (see Fig. 1.46):

- Minimalist Design.
- Strong Focus on Functionality.
- Balanced Proportions.
- Restrained Color Palette.
- Circular watch cases.
- Straight, thin hands.
- Simple index markers.
- Clear typography.

The Bauhaus movement was a German design school and artistic movement founded in 1919 that sought to unite art, craftsmanship, and technology. Its philosophy emphasized simplicity, functionality, and clean geometric forms, and the principle that *form follows function*.

Fig. 1.45 Wristwatch in Art Deco style

Today, Bauhaus continues to have a profound influence on architecture, industrial design, furniture, graphic design, and digital products. Many modern buildings, consumer products, and user interfaces reflect Bauhaus ideals through their minimalist esthetics, clarity, and focus on usability. The movement's emphasis on functional, timeless design remains a cornerstone of contemporary design practice.

A Bauhaus-style wristwatch is a watch whose design is inspired by the principles of the Bauhaus movement. This esthetic remains popular today because it gives watches a timeless, elegant appearance that avoids trends and excessive ornamentation.

Fig. 1.46 Wristwatch in Bauhaus style

Modernism (ca. 1920s to today)

Typical Elements of a Modernist-style watch (see Fig. 1.47):

- Radical break with historical styles and ornament.
- Emphasis on simplicity, clarity, and rationality.
- A sense of timelessness and universality in design.
- Watches designed to be both useful tools and esthetic statements.

Modernism was a major design and artistic movement that emerged in the early 20th century, driven by a desire to break from historical styles and embrace the possibilities of the modern industrial age. Characterized by simplicity, functionality, and the rejection of unnecessary ornamentation, Modernist design emphasized clean lines, geometric forms, and rational structures. It promoted the use of innovative materials and techniques, including steel, glass, and concrete, aiming to create designs that were both practical and aesthetically timeless. Influencing architecture, furniture, graphic arts, and industrial design, Modernism sought to harmonize form and function, reflecting the ideals of efficiency, clarity, and universal appeal, and leaving a lasting legacy on contemporary design worldwide.

Fig. 1.47 Wristwatch in Modernist style

Memphis Style (1981 to 2010)

Typical Elements of a Memphis Style watch (see Fig. 1.48):

- Bright, contrasting colors.
- Geometric shapes: Circles, triangles, zigzags, and asymmetrical forms.
- Cartoon-like, with inspiration from Art Deco and Pop Art.
- Founded in Milan in 1981 by Ettore Sottsass and a group of young
- designers and architects known as the Memphis Group.

The Memphis Style was an avant-garde design movement that emerged in the early 1980s, founded by the Italian designer Ettore Sottsass and the Memphis Group. It is characterized by bold colors, geometric shapes, playful patterns, and unconventional materials, often combining elements in surprising and whimsical ways. Rejecting the minimalism and functionalism of earlier movements, Memphis design embraced eccentricity, decorative excess, and a sense of humour, challenging traditional notions of taste and utility. Influencing furniture, lighting, textiles, and graphic design, the Memphis Style left a lasting impact on postmodern esthetics, celebrating creativity, individuality, and visual experimentation.

Fig. 1.48 Wristwatch in Memphis style

Low Poly Design (ca. 2010s to today)

Typical Elements of a Low Poly Design watch (see Fig. 1.49):

- Simple geometric shapes.
- Sharp edges and faceted surfaces.
- Minimalist, digital-inspired look.
- Sense of clarity, reduction, and futuristic esthetics.

Low Poly is a contemporary design style that emerged from 3D computer graphics and digital art, characterized by geometric simplicity, angular shapes, and faceted surfaces. Originally used in early computer graphics and video games to optimize rendering, it has evolved into an aesthetic choice that emphasizes minimalism, abstraction, and stylized forms. Low Poly design often features bold colors, clear edges, and simplified representations of objects or figures, creating visually striking compositions that balance technical precision with artistic creativity. This style has

Fig. 1.49 Wristwatch in Low Poly design style

influenced not only digital media but also illustration, animation, product design, and visual branding, celebrating clarity, efficiency, and modern digital esthetics.

1.3 Short History of Wristwatch Design

The history of wristwatch design is a story of utility, elegance, and innovation—shaped by social change, technological progress, and visionary designers. From its aristocratic origins to its place as a universal symbol of personal style, the wristwatch reflects both the spirit of its age and the imagination of those who crafted it.

Origins: The Queen of Naples and Early Wristwatches

The very first known wristwatch is credited to Caroline Murat, Queen of Naples, who in 1812 commissioned Abraham-Louis Breguet, the great watchmaker, to create a timepiece she could wear on her wrist. At the time, watches were overwhelmingly pocket watches, and wrist-worn timekeepers were regarded as feminine ornaments rather than practical instruments. For much of the 19th century, wristwatches remained a novelty, typically crafted as decorative jewelry for women.

The Birth of the Modern Wristwatch: Cartier and Santos-Dumont

The wristwatch's status changed dramatically in the early 20th century. One pivotal moment came in 1904, when pioneering aviator Alberto Santos-Dumont asked his friend Louis Cartier for a timepiece he could use while flying. Cartier responded by creating the Santos—a square, purpose-built wristwatch with exposed screws and a leather strap. Unlike the ornamental jewelry watches of the 19th century, the Santos was designed for functionality, readability, and style. It is often regarded as the first truly modern wristwatch design and remains an enduring icon today.

The Military Era: Function Shapes Form

The wristwatch truly entered the mainstream during World War I. Soldiers found pocket watches impractical in combat, leading to the widespread adoption of wristwatches as rugged, legible instruments. Wire lugs were added to pocket-watch cases, luminous paint improved visibility, and robustness became a design priority. By the 1920s, wristwatches were no longer dismissed as feminine but had become essential tools for men and women alike.

Mid-Century Elegance: The Age of Refinement

From the 1930s to the 1960s, wristwatch design matured into a spectrum of styles suited to different lives and professions. Dress watches emphasized slim, elegant cases and minimalist dials for formal wear. Dive watches, like the Rolex Submariner (1953) and Blancpain Fifty Fathoms (1953), were developed for professional divers, with rotating bezels and superior water resistance.

Pilot watches evolved into highly legible instruments with oversized crowns and luminous markers. Chronographs became popular with motorsport and aviation communities, blending technical utility with dynamic esthetics. In these decades, design was a balance of practicality and elegance, with stainless steel cases, domed crystals, and classic proportions defining much of the era.

In 1957, the Hamilton Ventura became the world's first electric watch powered by a battery. Beyond its technical breakthrough, the Ventura stood out for its avant-garde asymmetrical case design by Richard Arbib. Its futuristic look symbolized the optimism of the Space Age. It showed that watches could embody modernity and pop culture as much as tradition.

Times of Crisis: The Quartz Revolution

The 1970s brought seismic change: the quartz crisis. With Seiko's Astron (1969) and the mass production of quartz movements, watches became more accurate and affordable than ever. Traditional Swiss manufacturers were shaken, and design entered an era of bold experimentation.

Yet survival in this era also required fresh thinking at the mass-market level. In 1983, the Swatch Plastic Watch was launched by Nicolas G. Hayek and his team. With bold colors, playful graphics, and a lightweight plastic case, the Swatch redefined the wristwatch as a fashion accessory as much as a timepiece. Affordable and expressive,

Swatch watches became cultural icons of the 1980s and played a decisive role in saving the Swiss watch industry from collapse during the quartz crisis.

Contemporary Design: Tradition Meets Innovation

Since the 1990s, wristwatch design has reflected both heritage and modernity. Collectors and brands revived historical designs—reissues of military, pilot, and dive watches—while also pushing innovation with new materials like ceramics, titanium, and carbon composites. High-end watchmaking increasingly emphasized artistic finishing, open-worked dials, and ultra-complicated movements, blending technical mastery with visual artistry.

At the same time, the 21st century brought the rise of the smartwatch, beginning with the Apple Watch (2015) and similar models. Here, design emphasizes digital versatility: customizable faces, fitness functions, and seamless integration with daily technology. While far removed from traditional horology, smartwatches continue the story of the wristwatch as a practical tool adapted to its era.

The wristwatch has continually evolved in response to human needs and creativity. More than a device to tell time, it has become a canvas of design, expressing elegance, engineering, and identity. The story of wristwatch design is ultimately the story of how a small object on the wrist reflects the changing currents of culture, technology, and art [29, 30].

1.4 The Most Important Designers of Wristwatches

Among the many creative minds associated with horology, four designers stand out because of their long-lasting influence, originality, and cultural impact: Gérald Genta, Jorg Hysek, Max Bill, and Emmanuel Gueit. Each of them transformed the appearance of wristwatches in a unique way and introduced design principles that continue to shape contemporary watchmaking.

Gérald Genta: The Architect of the Luxury Sports Watch

No discussion of wristwatch design can begin with anyone other than Gérald Genta. Often described as the most influential watch designer of the twentieth century, Genta created some of the most recognizable wristwatches ever produced. His most famous works include the Audemars Piguet Royal Oak, the Patek Philippe Nautilus, the IWC Ingenieur SL, and the Universal Genève Polerouter.

Genta revolutionized watch design in the 1970s by introducing the concept of the luxury sports watch made of stainless steel. Before the Royal Oak appeared in 1972, luxury watches were generally thin, elegant dress watches made of precious metals. Genta broke this convention by designing a large steel watch with visible screws, an integrated bracelet, and an industrial esthetic inspired by a diver's helmet. The Royal Oak was initially controversial, but it eventually became one of the most successful luxury watches ever made.

His later design for the Patek Philippe Nautilus further refined this idea. Inspired by a ship's porthole, the Nautilus combined sporty robustness with refined elegance. These watches fundamentally changed the economics and esthetics of Swiss watchmaking and remain highly desirable today. Genta's genius lay in his ability to combine strong geometric forms with elegance and wearability. He also understood branding and visual identity better than most designers of his time [31].

Jorg Hysek: Sculptural Modernism in Watchmaking

Jorg Hysek represents the next generation of influential watch designers. Born in East Berlin and educated in Switzerland and London, Hysek brought a more sculptural and futuristic approach to horology. He worked for numerous luxury brands, including Vacheron Constantin, TAG Heuer, Breguet, Cartier, and Tiffany & Co.

One of Hysek's most important achievements was the design of the Vacheron Constantin 222 in 1977. This watch anticipated the style later associated with the Vacheron Constantin Overseas collection and became an important competitor to Genta's Royal Oak and Nautilus. Hysek also helped create TAG Heuer's Kirium collection in the 1990s, a design that reflected contemporary industrial esthetics and aerodynamic forms.

Unlike Genta, whose designs often emphasized bold geometry, Hysek explored fluidity and three-dimensionality. His watches frequently appear almost architectural, with smooth transitions between case and bracelet. This approach helped modernize watch design during the late twentieth century and influenced the emergence of more experimental luxury sports watches [32].

Max Bill: Bauhaus Principles on the Wrist

Max Bill was a Swiss architect, artist, and designer deeply associated with the Bauhaus tradition. His collaboration with the German watch manufacturer Junghans produced some of the purest examples of modernist watch design.

The Junghans Max Bill watches, first introduced in the 1960s, are celebrated for their simplicity, harmony, and mathematical precision. Every detail of these watches serves a functional purpose. The typography is clean and highly legible, the dials are uncluttered, and the proportions are carefully balanced.

Bill believed that industrial products should combine beauty and utility in equal measure. His watches embody the Bauhaus principle that form should follow function. Unlike luxury sports watches that communicate status and power, Max Bill watches communicate intellectual elegance and rational order.

The significance of Max Bill lies in his ability to transfer the ideals of modernist architecture and graphic design into wearable objects. His work demonstrated that wristwatches could function as expressions of philosophical and artistic ideas [33].

Emmanuel Gueit: The Designer Who Reinvented the Modern Luxury Watch

Among the most influential watch designers of the late twentieth century, Emmanuel Gueit occupies a special position. His designs helped redefine what a luxury wristwatch could look like in an era when the Swiss watch industry was struggling to adapt to changing tastes, technological disruption, and new cultural expectations.

Above all, Gueit is remembered as the creator of the revolutionary Audemars Piguet Royal Oak Offshore, one of the most important watches of the modern era.

Several characteristics distinguish Emmanuel Gueit's work from that of earlier watch designers. First, he emphasized scale. Before the 1990s, most luxury watches remained relatively compact. Gueit understood that larger watches created greater visual presence and emotional impact. His designs transformed the wristwatch into a bold statement piece rather than a subtle accessory. Second, Gueit embraced visible technicality. His watches often highlighted screws, gaskets, pushers, and structural elements that earlier designers might have attempted to hide. This approach reflected broader industrial design trends that celebrated engineering esthetics. Third, his work incorporated strong ergonomic and architectural forms. Cases appeared sculptural and muscular rather than merely decorative. The relationship between bezel, case, and bracelet became central to the visual identity of the watch.

Gueit understood the emotional power of luxury sports watches. His designs communicated energy, masculinity, and modernity in ways that resonated strongly during the 1990s and early 2000s. Although Emmanuel Gueit designed relatively few watches compared to some other famous designers, his influence remains extraordinary because of the cultural importance of the Royal Oak Offshore. Few individual watch designs have changed industry trends so dramatically [34].

1.5 Watch Design in The Future

The design of wristwatches is currently undergoing a profound transformation, shaped by technological innovation, shifting consumer values, and esthetic reinterpretations of the past. Looking ahead, several key developments are likely to define the future of watch design.

First, the convergence of traditional watchmaking and digital technology will continue to accelerate. Hybrid watches—combining analogue esthetics with smart functionalities—are expected to become increasingly dominant. Consumers are seeking devices that offer health tracking, connectivity, and AI-supported features without sacrificing the timeless elegance of classical designs. This trend reflects a broader movement toward "invisible technology," where digital capabilities are seamlessly integrated into familiar forms.

At the same time, design language is becoming more expressive. While minimalist styles remain relevant, bold colours and distinctive materials are gaining importance as tools of personal expression. Future watches are likely to feature vibrant dials, unconventional textures such as ceramic or meteorite, and even natural materials like stone. In parallel, monochromatic and matte finishes are emerging as a countertrend, emphasizing subtle sophistication over high-gloss luxury. Recent industry releases also confirm this shift toward colour experimentation and artistic dials, highlighting the growing importance of individuality in watch esthetics.

Another important development is the move toward slimmer and more ergonomic designs. Oversized watches are gradually being replaced by smaller case sizes and ultra-thin profiles that prioritize comfort and wearability. This reflects changing fashion trends and a preference for objects that integrate seamlessly into everyday life rather than dominate it [35].

Sustainability will also play a central role in shaping future watch design. Increasingly, manufacturers are adopting recycled metals, bio-based materials, and environmentally responsible production methods. This shift is driven by growing consumer awareness and the expectation that luxury products should also meet ethical and ecological standards [36].

In addition, personalization and modularity are expected to expand significantly. Interchangeable straps, customizable dials, and even modular components will allow users to adapt their watches to different contexts and styles. Watches are thus evolving from static accessories into dynamic, user-centered products.

Finally, experimental and conceptual approaches to time display may redefine how watches are perceived. New designs are already exploring alternative ways of presenting time—sometimes even abandoning traditional hands—while integrating artistic and sensory elements. Such innovations suggest that watches may increasingly function not only as instruments but also as experiential objects that blend design, technology, and emotion [37].

In conclusion, the future of wristwatch design lies in the balance between heritage and innovation. Watches will likely become more technologically sophisticated, environmentally responsible, and individually expressive, while still preserving the emotional and cultural value that has defined them for centuries.

1.6 How to Tell a Watch Story

Watches do not exist in isolation. They are born into a cultural, historical, and emotional context, and they draw much of their meaning from the stories they tell. Within the watch industry, storytelling has become one of the most powerful tools for creating identity, value, and longevity. A well-designed timepiece not only measures the passing hours but also communicates a narrative—of heritage, craftsmanship, innovation, or personal aspiration.

From the earliest days of horology, watchmakers recognized that precision alone could not sustain fascination. Technical mastery was essential, but it was the narrative surrounding a watch that captured the imagination. Stories of pioneering explorers, astronauts, or divers who relied on their timepieces in extreme conditions transformed watches into symbols of adventure.

Cartier offers one of the earliest and most enduring examples. When the pioneering aviator Alberto Santos-Dumont asked Louis Cartier to design a practical wristwatch for flying, the result was the Cartier Santos—one of the first men's wristwatches. The story of friendship between inventor and craftsman, fused with the romance of early aviation, became a cornerstone of Cartier's identity: refined elegance with a spirit of progress. Similarly, Rolex built its legend through real feats of endurance. In 1927, Mercedes Gleitze swam the English Channel wearing a Rolex Oyster, proving its waterproof design. The publicized success established Rolex as the watch for pioneers and achievers. This narrative deepened in 1953, when Sir Edmund Hillary and Tenzing Norgay conquered Mount Everest with Rolex Oyster Perpetual watches on their wrists. These authentic moments of exploration laid the foundation for Rolex's timeless slogan: A crown for every achievement.

TAG Heuer crafted its mythology through cinema and speed. When Steve McQueen wore the bold Heuer Monaco in the 1971 film Le Mans, he personified the brand's racing spirit—rugged, daring, and effortlessly cool. Decades later, McQueen remains one of the most iconic faces of TAG Heuer, proving the lasting power of cultural storytelling.

For Omega, the connection between storytelling and fiction reached a global scale through James Bond. Since 1995's Golden Eye, 007 has worn the Omega Seamaster, merging cinematic fantasy with technical sophistication. Bond's character—fearless, elegant, and innovative—perfectly mirrors Omega's identity, and every new film reinforces this association between heroism and horology.

Other brands have also built powerful narratives through real-life pioneers. Longines celebrates Amelia Earhart, who wore the brand on her solo flight across the Atlantic in 1932, symbolizing courage and precision. Breitling entered the space age with Scott Carpenter, whose Navitimer Cosmonaute orbited Earth in 1962, linking the brand to both aviation and exploration. Meanwhile, Panerai transformed from a supplier to the Italian Navy's elite divers into a global luxury brand after Sylvester Stallone famously wore a Panerai Luminor in the film Daylight. His spontaneous endorsement turned a once-secret military tool into a Hollywood status symbol.

If storytelling is central to the industry, it is equally vital at the level of design. Every choice a designer makes—shape, material, texture, typography—contributes to the narrative of the watch. A tool watch with a brushed steel case and oversized numerals speaks of resilience and practicality, evoking the needs of aviators or divers. A slim gold case with Roman numerals tells of refinement, heritage, and ceremonial elegance. Even contemporary minimalist designs tell a story, often one of restraint, clarity, and modernity.

This process is not merely decorative. Storytelling in design is about coherence: ensuring that every visual and tactile element supports the narrative. If a brand claims to build watches for explorers, the design must feel purposeful, rugged, and legible in any condition. If a watch is meant to evoke a golden age of artistry, its finishing, proportions, and details must whisper of craftsmanship and permanence.

Successful storytelling in design often draws from three sources:

Heritage

Many designs reference the archives, echoing the shapes, hands, or dials of historical models. This continuity tells the story of a lineage that has been preserved.

Purpose

Function-driven design communicates credibility. The clarity of a diver's bezel or the strength of a pilot's crown tells a story of trust in demanding circumstances.

Emotion

Beyond heritage and purpose, design can evoke mood: elegance, adventure, modernity, rebellion. It creates a resonance that speaks directly to the wearer's sense of identity.

Perhaps the most powerful storytelling happens not in the brand's narrative, but in the wearer's own. A watch becomes part of a personal story the moment it is chosen: a gift marking an achievement, a reminder of a journey, or a daily companion that quietly witnesses the passing of life. Designers who understand this dimension create watches that invite stories—pieces that feel timeless, adaptable, and capable of carrying meaning through generations.

Storytelling is not an addition to watch design—it is its lifeblood. Without narrative, even the most precise movement risks becoming irrelevant in a digital age. With it, a watch transcends utility, becoming an object of culture, memory, and aspiration. To design a watch, then, is not only to shape form and function, but to craft a vessel for stories—stories of brands, of history, and ultimately of people who find a fragment of themselves reflected in the language of time.

1.7 How to Create a Luxury Watch

A watch is far more than a tool for telling time; it is an emblem of taste, craftsmanship, and status. While countless watches can be purchased for practical use, only a select few achieve the rarefied category of luxury. The making of a luxury watch is not merely a matter of adding expensive materials—it is the harmonious blending of design, artistry, engineering, and heritage. To understand how such timepieces are created, one must first consider what gives a watch its aura of luxury and then examine how skilful design elevates it into an object of enduring value.

The aura of luxury begins with perception. A luxury watch signals refinement, exclusivity, and discernment. This perception often arises from several intertwined factors: heritage, rarity, craftsmanship, and material excellence.

Heritage and brand legacy: Many luxury watches are tied to long-standing traditions. A brand with centuries of innovation conveys trust and authority, and its products carry the weight of this history.

Exclusivity

Limited editions and bespoke models foster desirability. A watch that is not easily accessible becomes a symbol of distinction.

Craftsmanship

The visible and invisible labour of skilled artisans imbues each piece with authenticity. Hand-finished movements, intricate complications, and perfectly polished cases all suggest care and mastery.

Materials

The use of precious metals, high-grade sapphire crystals, and innovative alloys distinguishes luxury watches from mass-market alternatives. Even the strap—crafted from fine leather or advanced ceramics—contributes to this aura.

Luxury, therefore, is not just an attribute but a story told through the watch's origin, composition, and rarity.

Design is the most immediate communicator of luxury. Before one learns about the movement inside or the materials used, the eye judges the harmony of form. A truly skilful design goes beyond esthetics; it subtly conveys value through balance, detail, and innovation.

Proportions and harmony

A watch with well-considered proportions radiates refinement. The interplay between dial, bezel, and case must feel intentional, with no element appearing forced or excessive.

Attention to detail

Luxury design reveals itself under close inspection. Applied indices, hand-polished hands, guilloché patterns, and bevelled edges give the watch a depth that cannot be replicated by mass production.

Innovation with restraint

Complications such as moon phases, tourbillons, or perpetual calendars showcase technical mastery, but they must be integrated seamlessly into the design. True luxury avoids clutter, maintaining elegance even when complexity is present.

Emotional resonance

Design is not only visual but also tactile. The smooth action of a crown, the satisfying click of a clasp, and the weight of the watch on the wrist create a sensory experience of quality.

Through design, a watch conveys a sense of permanence. When form and function merge flawlessly, the owner perceives not just an accessory but a piece of art—a small sculpture that marks the passage of time.

Creating a luxury watch is a delicate orchestration of heritage, materials, craftsmanship, and design. While the movement inside may be a marvel of engineering, it is the aura of luxury—communicated through story and artistry—that transforms a watch into a coveted treasure. Skilful design amplifies this impression, ensuring that every glance at the dial confirms the wearer's investment in value and elegance. In the end, a luxury watch is not just made—it is crafted, with every detail serving to remind its owner that true luxury resides in things that transcend utility.

1.8 How to Become a Watch Designer

Many people dream of designing wristwatches professionally. Their goal is either to work as a full-time designer for a renowned watch manufacturer or as a freelancer for various producers. The path to achieving this goal is not clearly defined. Essential requirements include technical understanding, a penchant for meticulous work, and a sense of esthetics. While not required, a passion for wristwatches is certainly helpful.

Ideally, anyone applying for a watch design position should have completed training as a product, industrial, or jewelry designer. Graduates of precision engineering programs with a compelling portfolio of their own watch designs also have good chances.

A particularly promising option is to visit one of these schools with a high reputation in the luxury industry:

Creative Academy (Richemont Creative Academy Milan)

The Creative Academy is the corporate design school of the luxury group Richemont, owner of prestigious watch brands such as IWC Schaffhausen, Jaeger-LeCoultre, Vacheron Constantin, Panerai, Piaget, and Cartier. The school was founded in 2003 specifically to train young designers for careers in luxury watches, jewelry, and accessories. Each year, only about twenty students from leading international design schools are admitted. The school strongly emphasizes "learning by doing." Students work on real projects directly supervised by Creative Directors and executives from Richemont brands. For aspiring watch designers, Creative Academy is considered one of the clearest direct pathways into the Swiss luxury-watch sector [38].

ECAL

ECAL (École cantonale d'art de Lausanne) is one of Europe's most respected art and design universities. The school is internationally known for industrial design, luxury product design, photography, and visual communication. Many graduates later work in the Swiss luxury industry, including watchmaking and jewelry.

HEAD—Genève (Geneva University of Art and Design)

HEAD is a leading Swiss art and design school with strong links to fashion, jewelry, and luxury industries. Because Geneva is closely connected to the Swiss watch sector, the school has valuable industry connections.

Politecnico di Milano

Politecnico di Milano is one of Europe's most prestigious technical universities. It is especially famous for industrial and product design. Italy's strong tradition in fashion, automotive, and luxury design strongly influences the programs.

Birmingham City University School of Jewellery

The School of Jewellery is part of Birmingham City University and is in Birmingham's historic Jewelry Quarter. It is internationally respected for jewellery, metalworking, and accessory design education.

WOSTEP

WOSTEP (Watchmakers of Switzerland Training and Educational Program) is one of the world's most respected watchmaking schools. Unlike art-focused design schools, WOSTEP specializes in professional horology and technical watchmaking education.

Institut Supérieur de Design

This French design institute is known for industrial and transportation design, but also prepares students for the luxury product industries through strong conceptual and technical design education.

École Boulle

École Boulle is one of France's historic elite schools for decorative arts, fine craftsmanship, jewelry, and luxury object design.

La Haute École de Joaillerie

Originally connected to the French jewelry industry, this institution specializes in luxury jewelry creation and gem-related education.

NAIOSW

The North American Institute of Swiss Watchmaking (NAIOSW) is associated with Richemont technical watchmaking education in the United States. Unlike Creative Academy, this institution focuses primarily on technical watchmaking rather than artistic design.

The institutions mentioned above share several important educational priorities, even though some focus more on artistic design while others emphasize technical horology or craftsmanship. Together, they reflect the multidisciplinary nature of modern wristwatch creation. Despite their differences, all of these institutions share one central philosophy:

A successful watch designer must combine artistic sensitivity, technical understanding, craftsmanship, and professional discipline.

Exercise 1

Solve the multiple-choice quiz. Find the correct completion for each sentence. Only one is correct in each case. You'll find the solutions at the end of the book.

1.1 The oscillator's function is to ...

- a ☐ ... to control the escapement inside a watch movement.
- b ☐ ... to wind up the watch movement automatically.
- c ☐ ... to give power to the gear train of a watch movement.

1.2 The Japanese company Seiko was the first to develop...

a ☐ … a watch with a digital display.
b ☐ … a tuning fork movement for watches.
c ☐ … a quartz movement for watches.

1.3 A regulateur watch…

a ☐ … regulates itself by receiving a radio signal.
b ☐ … shows the time in multiple time zones simultaneously.
c ☐ … separates the hour, minute, and second hands onto different sub-dials.

1.4 The tourbillon…

a ☐ … was invented in the 14th century by Giovanni de Dondi.
b ☐ … was invented in 1801 by Abraham-Louis Breguet.
c ☐ … was invented in 1821 by Nicolas Mathieu Rieussec.

1.5 UTC stands for…

a ☐ … Universal Time Chronograph.
b ☐ … Universal Time Coordinated.
c ☐ … Universal Time Chronometer.

1.6 Regatta watches allow sailors...

a ☐ … to track the crucial interval before a race starts.
b ☐ … to determine on a long-distance regatta their east–west position at sea.
c ☐ … to estimate the distance to a visible and audible event.

1.7 The Memphis design style was developed in…

a ☐ … Memphis.
b ☐ … Munich.
c ☐ … Milan.

1.8 Alberto Santos Dumont, friend of Louis Cartier, was a famous…

a ☐ … race driver.
b ☐ … aviator.
c ☐ … architect.

1.9 The Patek Philippe Nautilus was designed by…

a ☐ … Gérald Genta.
b ☐ … Emmanuel Gueit.
c ☐ … Jorg Hysek.

1.10 In the movie “Daylight,” Sylvester Stallone wore a…

a ☐ … IWC Aquatimer.
b ☐ … Panerai Luminor.
c ☐ … Omega Seamaster.

Chapter 2
Toolbox of Watch Designers

Abstract Chapter 2 presents the "toolbox" available to watch designers, examining the principal elements and design possibilities of a wristwatch. It covers case shapes, materials, finishing techniques, proportions, construction, crystals, case backs, crowns, and gemstone settings. The chapter then explores dial design, including surface treatments, printing, proportions, colour, typography, indices, luminosity, layouts, and specialized scales. It also addresses hands, straps, bracelets, clasps, and attachment systems, before examining movement aesthetics, bridge forms, finishing, engravings, rotors, and quality seals. Finally, it discusses personalization, emphasizing how customization, craftsmanship, functionality, and emerging materials enable distinctive and individual timepieces.

Just as skilled craftsmen find the right tool for every problem in their toolbox, watch designers have numerous design tools at their disposal. The second part of this book explains the most important design elements of a wristwatch and their possible variations.

The toolbox described is vast, virtually endless. Besides the tools outlined in this chapter, it offers ample space for previously unknown elements. It is the task of every watch designer to constantly seek out new techniques and elements suitable for watches. These might include manufacturing methods that allow innovative, exciting shapes. Materials are equally interesting. Research in plastics and metallurgy regularly reports on new developments. Nanotechnology makes it possible to give dial surfaces that were unproducible in the past. And as for watch straps, the growing range of sustainable materials should not be overlooked.

G. Plange, *A Tutorial for Watch Designers*,
https://doi.org/10.1007/978-3-032-33722-1_2

2.1 Case Design

2.1.1 Case Shapes

When it comes to the shape of a watch case, there are no limits to designers' creativity. A watch case should, above all, be comfortable to wear and easy to produce. Furthermore, it should stand out aesthetically from those of its competitors. A unique, innovative design helps signal a watch's lineage from a particular manufacturer.

The shapes of watch cases can be roughly divided into the groups shown in Fig. 2.1 [39].

2.1.2 Case Materials

Numerous different materials are used for watch cases. They differ in terms of their value, weight, and the effort required for processing. The following materials are the most commonly used [40].

Stainless Steel

- Properties: Durable, corrosion-resistant, relatively affordable.
- Benefits: The most common watch material due to its strength, scratch resistance, and versatility. It offers a good balance between durability, weight, and cost, making it suitable for everyday wear.

Titanium

- Properties: Lightweight, strong, hypoallergenic, corrosion-resistant.
- Benefits: Much lighter than steel but equally robust, making it comfortable for larger watches. Its hypoallergenic nature makes it ideal for sensitive skin. Often used in sports and dive watches.

Tantalum

- Properties: Dense, heavy, bluish-grey metal with excellent corrosion resistance.
- Benefits: Gives watches a unique, luxurious feel due to their weight and rare appearance. It is extremely resistant to scratches and tarnishing, appealing to collectors who value exclusivity.

Bronze

- Properties: Durable, develops a natural patina over time.
- Benefits: Popular in dive watches, bronze cases age uniquely, creating a vintage look. The patina layer also protects the metal from further corrosion, giving each watch a distinct character.

Aluminium

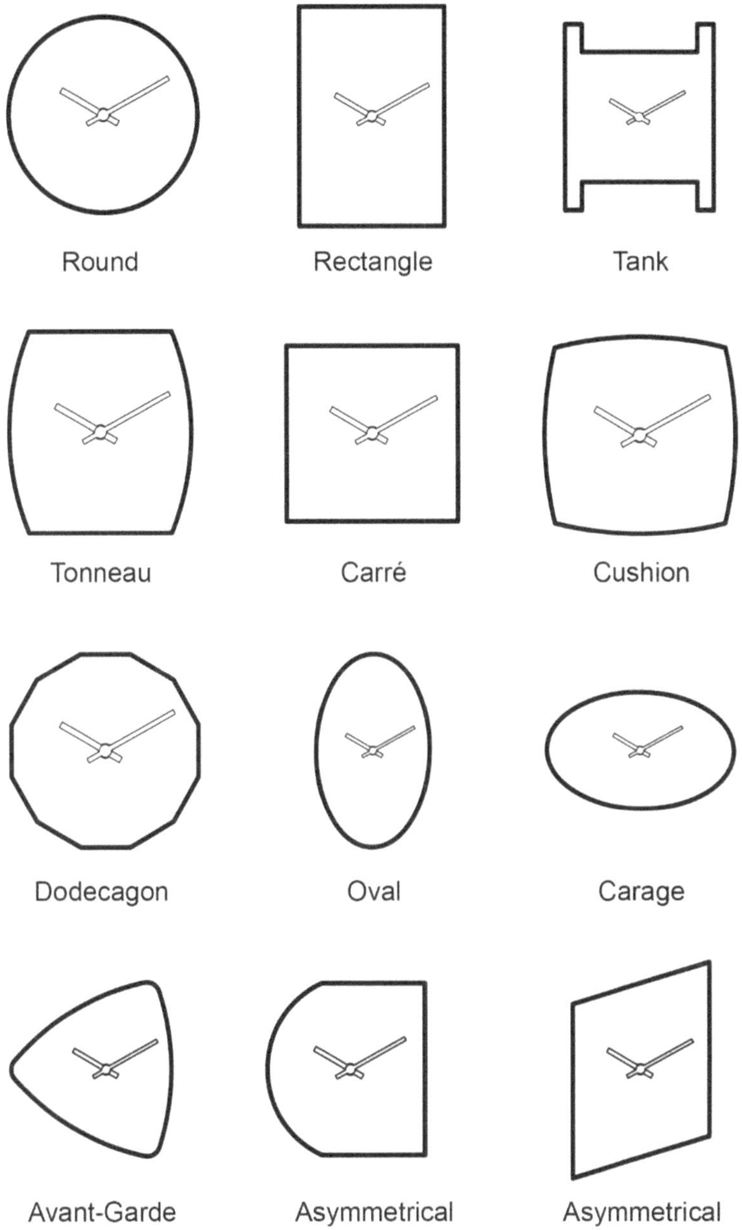

Fig. 2.1 Traditional watch case shapes

- Properties: Very lightweight, can be anodized in different colors.
- Benefits: Used for sporty and colourful designs. While less durable than steel or titanium, its lightness makes it extremely comfortable on the wrist.

Gold (Yellow, White, Rose)

- Properties: Precious, soft metal, resistant to corrosion, luxurious appearance.
- Benefits: Associated with prestige and elegance. Gold watches are more decorative than tool-oriented, often used in high-end dress watches. Alloys (rose or white gold) increase hardness and add different colors.

Platinum

- Properties: Very dense, heavy, rare, and highly resistant to corrosion.
- Benefits: Considered the ultimate luxury metal. Its weight gives a sense of solidity and exclusivity. Platinum watches are discreet yet extremely valuable, appealing to connoisseurs.

Plastic/Resin

- Properties: Lightweight, inexpensive, impact-resistant.
- Benefits: Common in digital and sports watches. Provides excellent shock resistance and comfort, making it ideal for outdoor and active lifestyles.

Sapphire (Synthetic Sapphire)

- Properties: Extremely hard and scratch-resistant, transparent.
- Benefits: Mostly used for crystals, but some brands produce sapphire cases. A sapphire case is nearly scratch-proof and provides a unique, see-through esthetic, though it is brittle compared to metals.

Ceramics

- Properties: Very hard, scratch-resistant, lightweight, hypoallergenic.
- Benefits: Provides a glossy, luxurious appearance that doesn't fade. Unlike metal, ceramic does not scratch easily, though it can shatter if dropped. Often used in bezels or entire cases for a modern look.

Carbon Fiber/Forged Carbon

- Properties: Extremely lightweight, strong, modern composite material.
- Benefits: Offers strength without weight, perfect for high-performance or sporty watches. Its unique layered or marbled texture gives a technical, futuristic esthetic.

Figure 2.2 shows the main differences between the materials discussed above.

Material	Ligation	Density (g/cm³)	Melting Point (°C)	Hardness (Vickers)	Can cause Allergies
Stainless Steel	Iron + Carbon	7.84	1400 – 1450	100 – 200	Yes
Titanium	- (Element)	4.54	1668	200 – 400	Unlikely
Tantalum	- (Element)	16.6	3017	60 - 120	No
Bronze	Copper + Tin	8.73	930	50 - 300	Yes
Aluminium	- (Element)	2.7	660	15 - 50	Yes
Gold 750	Gold + Nickel	19.3	1064	140	Unlikely
Platinum 950	- (Element)	21.45	1768	150	No
Plastic	-	1.2 – 1.6	100 - 300	10 - 150	Possible
Ceramics	- (Molekule)	5.68	1768	1200	No

Fig. 2.2 Table of technical properties of common watch case materials

2.1.3 Case Surface Treatment

The surfaces of wristwatch cases can be refined through various finishing techniques, each giving the watch a distinct look, texture, and character. Common methods include [41].

Polishing

Process: Carried out using rotating polishing wheels made of cloth, felt, or leather, charged with abrasive pastes of different grit sizes.

- Produces a smooth, mirror-like finish with high reflectivity.
- Enhances elegance and highlights curves and edges.
- Often used on luxury watches, though it is prone to showing scratches.

Brushing/Satin Finishing

Process: Done using abrasive belts, Scotch-Brite wheels, or handheld abrasive pads.

- Creates a matte effect with fine, directional lines on the surface.
- Reduces glare and makes scratches less noticeable.
- Variations: Vertical, horizontal, circular, or radial brushing.
- Commonly applied to tool watches and sports models for a more understated look.

Sandblasting/Bead-Blasting

Process: The case is placed in a blasting chamber where compressed air propels fine particles (such as sand, glass beads, or ceramic beads) against the surface.

- Achieves a uniform, matte texture by bombarding the surface with fine particles.
- Provides a rugged, non-reflective appearance.
- Often used for military-style or technical timepieces.

Coating (PVD/DLC/Powder Coating)

Process PVD and DLC: The case is placed in a vacuum chamber. Metal vapours (e.g., titanium nitride) are deposited onto the case through ion bombardment.

Process Powder Coating: Involves electrostatically applying powdered paint, then curing it under heat:

- Adds a thin layer of coloured or protective material to the case (often black or gold).
- Increases scratch resistance and allows for a wide range of esthetic effects.
- PVD (Physical Vapor Deposition) and DLC (Diamond-Like Carbon).

Figure 2.3 shows some of the colors in which a PVD coating is possible.

Engraving

Process Hand Engraving: Done with small chisels (gravers) by skilled artisans.

Process Machine Engraving: CNC machines or pantographs can reproduce precise logos or decorative patterns.

Process Laser Engraving: Modern technique using high-powered lasers to etch text, logos, or textures with extreme precision.

- Decorative patterns, textures, or logos can be cut into the case surface.

Fluting

Process: Creation of parallel grooves, ridges, or fluted patterns on the watch case, bezel, or crown, either by milling, stamping, or hand engraving.

- It is often used decoratively (e.g., on classic fluted bezels) but can also provide better grip.
- The structured surface enhances both the tactile and visual character of the case.

Combination Finishes

Process: multiple finishing stages. For example, lugs may be brushed, bezels polished to a mirror finish, and case sides bead blasted.

- Many high-end watches combine polished, brushed, and matte surfaces on different areas of the case.
- This contrast emphasizes shapes and edges, creating visual depth and sophistication.

2.1.4 Soft Iron Inner Case

A soft iron cage inside a wristwatch case is used to protect the movement from magnetic fields. Magnetic fields can severely affect the accuracy of a mechanical watch by magnetizing its delicate components, especially the balance spring. The

Fig. 2.3 PVD coating colors

soft iron material can attract and redirect magnetic forces away from the movement, essentially creating a protective shield.

The main advantage of a soft iron cage, as presented in Fig. 2.4, is that it ensures reliable timekeeping even in environments with strong magnetic exposure, such as near electronic devices, industrial machinery, or medical equipment. This feature is particularly valuable for professionals working in technical, aviation, or scientific fields, where magnetic interference is common.

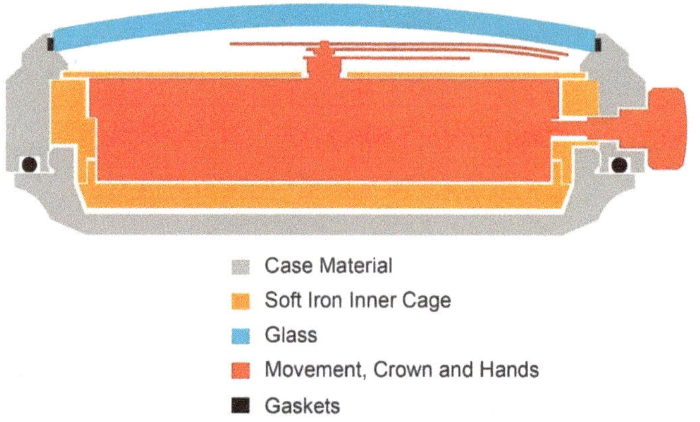

Fig. 2.4 Cross-section of a wristwatch with soft iron inner case

In addition, a soft iron cage enhances the durability and robustness of the watch, making it part of the tool-watch category. While modern anti-magnetic materials for movements are becoming more common, the soft iron cage remains a proven and trusted solution in traditional watchmaking [42].

2.1.5 Case Proportions

The ideal relationship between the diameter and the height, or thickness, of a wristwatch is widely regarded as a key factor in achieving visual balance, comfort, and wearability. In general, a well-proportioned watch should appear neither excessively flat nor overly bulky when viewed from the side, with its thickness harmonising naturally with its case diameter. As a commonly accepted guideline, the height of a watch case should amount to roughly one quarter to one third of its diameter. This ratio allows the watch to maintain a solid presence on the wrist while still sliding comfortably under a shirt cuff.

The lug-to-lug distance is a decisive geometric parameter that determines how the watch spans the wrist. Excessive lug length relative to wrist width results in overhang, reduced comfort, and compromised stability. Therefore, the lug-to-lug dimension should remain within the physical width of the wrist.

Beyond pure measurements, the perceived thickness of a watch is also influenced by design elements such as case shape, bezel width, crystal profile, and case back construction. Domed crystals, stepped cases, or tapered sides can make a thicker watch appear slimmer, while flat, vertical case walls can exaggerate height. Ultimately, the ideal diameter-to-height ratio ensures that a watch feels stable, comfortable, and aesthetically cohesive, reinforcing the impression of thoughtful and refined design [43, 44].

Fig. 2.5 Table of watch diameters for men and women

Diameter	Size (Sex)
26 -33	Small (Women)
34 - 36	Midsize (Women)
37 – 40	(Unisex)
41 - 42	Midsize (Men)
43 – 44	Large (Men)
45 +	Extra-large (Men)

Figure 2.5 shows the usual designations for men's and women's watches in different diameters.

2.1.6 Case Construction

Dress Watches

Their construction focuses on slimness, refinement, and visual harmony rather than robustness. Cases are typically thin and compact, often made from precious metals such as gold or platinum, and designed to slide easily under a shirt cuff. Water resistance is usually minimal, just enough to protect against incidental moisture. Crystals are often domed sapphire or mineral glass to enhance elegance, and movements tend to be slim mechanical calibres or high-quality quartz movements. Decorative elements such as polished surfaces, fine dial finishes, or gemstone settings are common, while protective features like crown guards or shock resistance play only a minor role.

Sports Watches

Robustness balances with wearability, making them suitable for active lifestyles without being tied to a single professional purpose. Cases are generally thicker and stronger than those of evening watches and are commonly made from stainless steel, titanium, or modern composites. Sports watches usually offer increased water resistance, reinforced cases, and sapphire crystals for scratch protection. Features such as screw-down crowns, luminescent hands and markers, and shock-resistant movements are typical. The overall design prioritises legibility and reliability, while still allowing for a wide range of styles and complications.

Divers' Watches

They are engineered to function reliably under significant water pressure and in low-visibility underwater conditions. As a result, their cases are usually thick and heavily reinforced, with high water resistance ratings, often starting at 200 m or more. Screw-down crowns and case backs are standard, and gaskets are specially designed to maintain tight seals under pressure. Some deep-diving watches include a helium escape valve. This doesn't keep water out directly but prevents internal pressure buildup during deep dives, which could otherwise damage the seals. A unidirectional rotating bezel with a clear minute scale is a defining feature, allowing divers to track immersion time safely. High-contrast dials, strong luminous material, and anti-reflective sapphire crystals ensure legibility in dark or murky water. Compared to other watch types, divers' watches prioritise safety, durability, and functional clarity over slimness or decorative refinement.

In watchmaking, the terms one-piece, two-piece, and three-piece watch case describe how the watch case is constructed from separate components. The construction method affects [45]:

- water resistance
- structural rigidity
- servicing complexity
- manufacturing cost
- overall design flexibility

One-Piece Case

A one-piece case—also called a monocoque case—is manufactured from a single main body. In this construction, the middle case and case back are integrated into one solid component. There is no removable case back; the movement is usually inserted from the front side of the watch. This construction became especially popular in professional dive watches and military watches.

Advantages:

- Very high structural rigidity
- Excellent water resistance
- Fewer sealing surfaces and gaskets
- Good shock resistance

Disadvantages:

- More difficult servicing
- The crystal and bezel often must be removed to access the movement
- Limited design flexibility
- Transparent display backs are usually impossible

Two-Piece Case

A two-piece case consists of the main case body (mid-case) and a removable case back. The movement is normally inserted from the back. Today, this is probably the most widespread construction type.

Advantages:

- Easier maintenance and servicing
- Simpler manufacturing

- Good balance between robustness and practicality
- Common in modern wristwatches

Disadvantages:

- Additional sealing surface between case and case back
- Slightly lower structural rigidity than monocoque cases

Three-Piece Case

A three-piece case separates the watch case into the case middle (mid-case), the bezel and the case back. The bezel is an independent component that secures the crystal and can sometimes rotate (for example, on dive watches). Three-piece constructions are common in luxury watches, sports watches and watches with complex case geometries.

Advantages

- High design flexibility
- Easier replacement of components
- Supports rotating bezels
- Allows sophisticated finishing and modular construction

Disadvantages

- More complex manufacturing
- More sealing interfaces
- Potentially more leak points

2.1.7 Watch Glasses

The "crystal" or "glass" of a wristwatch is the transparent cover that protects the dial and hands. It plays a crucial role not only in esthetics but also in durability, clarity, and scratch resistance. Over time, the watch industry has developed and refined several types of crystal materials, each with its own characteristics [46].

Acrylic (Plexiglass or Hesalite)

Acrylic is a type of transparent plastic, also known as Plexiglass or, in Omega's terminology, Hesalite. It was widely used in watches from the mid-20th century and is still found in some modern and vintage-inspired models.

Advantages:

- Very lightweight and inexpensive.
- Highly impact resistant; it won't shatter easily if dropped.
- Easily polished to remove scratches.

Disadvantages:

- Scratches easily compared to other materials.
- Less transparent and tends to age over time (yellowing or fine cracks).

Mineral Glass

Mineral glass is a type of hardened glass made from silica. It is heat-treated or chemically strengthened to improve its durability and is the most common material in mid-range watches.

Advantages:

- More scratch-resistant than acrylic.
- Affordable to produce and replace.
- Provides good optical clarity.

Disadvantages:

- Can shatter or crack upon strong impact.
- Once scratched, it cannot be polished—it must be replaced.

Sapphire Crystal

Synthetic sapphire crystal is made from crystallized aluminium oxide, grown in laboratories and then polished to optical clarity. It is the hardest and most premium watch crystal material used today.

Advantages:

- Extremely scratch-resistant (9 on the Mohs hardness scale; only diamond is harder).
- Excellent transparency and brilliance.
- Can be treated with anti-reflective coatings for improved legibility.

Disadvantages:

- More expensive to produce and replace.
- Brittle under strong impact—can chip or shatter if hit at the right angle.
- Anti-reflective coatings can scratch even if the sapphire itself does not.

Hardlex and Other Proprietary Materials

Some manufacturers use proprietary hardened glass types that combine the benefits of mineral and sapphire. The most famous example is Seiko's Hardlex, a specially treated mineral crystal with increased toughness.

Advantages:

- More impact-resistant than sapphire.
- Better scratch resistance than standard mineral glass.
- Cost-effective balance between durability and price.

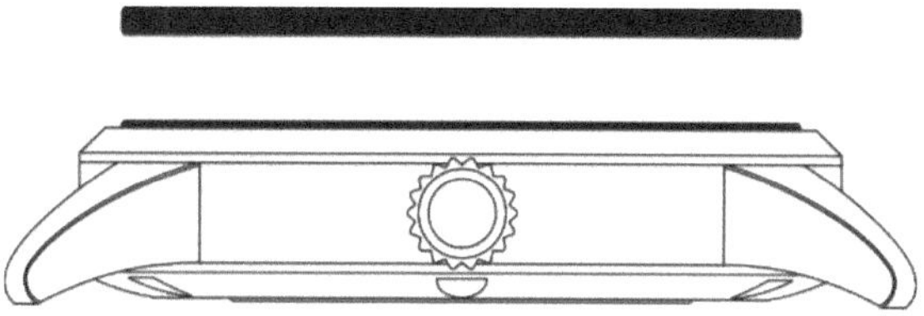

Fig. 2.6 Side view of wristwatch with flat watch glass

Disadvantages:

- Not as scratch-proof as sapphire.
- Typically brand-specific, so replacements must come from the manufacturer.

Here's a clear breakdown of the differences between flat, convex, and domed watch glasses.

Flat Watch Glasses

A flat watch crystal (see Fig. 2.6) has no curvature whatsoever. It appears clean, modern, and clear, and allows for a distortion-free view of the dial from almost any angle. Due to its simple geometry, it is relatively inexpensive to manufacture and replace. Furthermore, anti-reflective coatings can be applied particularly evenly. Disadvantages include a slightly higher tendency to reflect light in unfavourable conditions and lower shock resistance compared to domed crystals. Flat crystals are therefore frequently found in modern dress watches, field watches, and tool watches.

Convex Watch Glasses

A convex watch crystal (see Fig. 2.7) has a slight curvature and can be curved either only on the outside or on both the outside and inside. Single-curved crystals are flat on the inside, while double-curved versions are curved on both sides. Visually, they appear more elegant and softer than flat crystals and blend more harmoniously into the case. They offer slightly higher shock resistance and often sit more flush under a shirt cuff. Depending on the design, however, slight distortions can occur at the edges, especially with simply domed crystals. Manufacturing them is more complex and therefore more expensive than flat crystals, which is why they are frequently used in higher-end sports or dress watches.

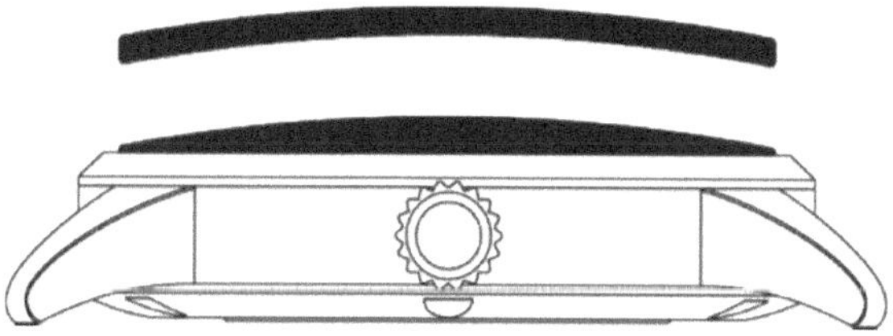

Fig. 2.7 Side view of wristwatch with convex watch glass (double curved)

Fig. 2.8 Side view of wristwatch with domed watch glass

Domed Watch Glasses

A domed or highly curved watch crystal (see Fig. 2.8) protrudes significantly beyond the bezel and strongly influences the appearance of a watch. They are reminiscent of classic vintage models and, depending on the viewing angle, create a distinct optical depth as well as a visible distortion of the dial. Technically, they offer very good shock resistance, as impact forces are distributed more effectively. However, this is offset by increased susceptibility to reflections and higher production costs. Furthermore, such crystals are more susceptible to contact with external surfaces and can be bothersome under tight cuffs. They are frequently used in vintage reissues or historical chronographs.

2.1.8 Watch Backs

As shown in Fig. 2.9, watches can have either a closed case back or an open case back, and the difference mainly lies in design, functionality, and the way the movement is presented [47].

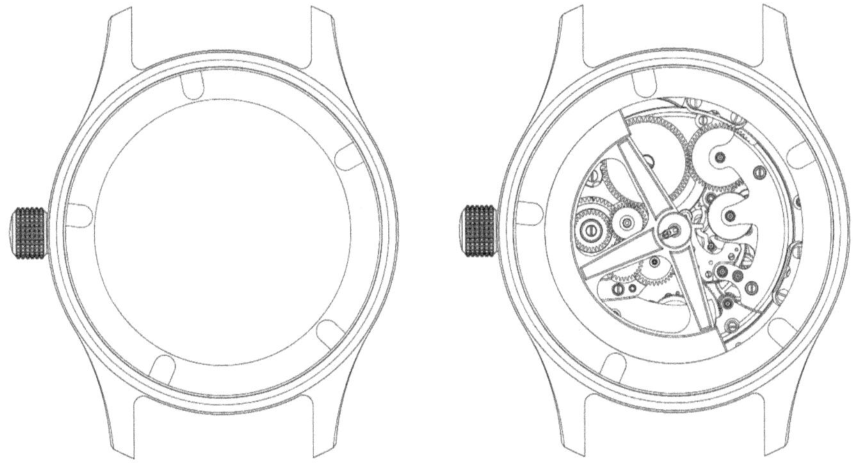

Fig. 2.9 Closed and open watch backs

Closed

A closed case back is made of solid material, usually stainless steel, titanium, or gold, and completely covers the movement. This design offers better protection against dust, moisture, and physical impacts. It is often preferred for tool watches, sports watches, and professional diving watches because it can improve durability and water resistance. A closed case back also allows space for engravings, such as serial numbers, logos, or commemorative text.

Open

An open case back features a transparent window, typically made of sapphire crystal or mineral glass, that allows the wearer to see the movement inside the watch. This design highlights the craftsmanship, decoration, and mechanical complexity of the movement. Open case backs are especially popular in mechanical and luxury watches, where visual appreciation of the movement is an important part of the experience. However, they may offer slightly less protection and sometimes reduced water resistance compared to fully closed case backs.

There are various methods for attaching watch backs to the case. The most common ones are listed below [48].

Snap-On/Press-Fit Case Backs

These case backs are pressed into place with a tight friction fit around the edge of the case. They do not use screws or threads—instead, they rely on precise tolerances so the back stays in place by pressure alone.

Advantages:

- Quick to assemble/disassemble.
- Lower manufacturing cost.

Disadvantages:

- Lower water resistance than screw types.
- Can be harder to remove without proper tools.

Screw-Down/Threaded Case Backs

The case back has threads around its edge and is screwed onto matching threads inside the case. This method yields a tight and secure seal, especially with a gasket, and is often found on dive and sport watches.

Advantages:

- High resistance to water and dust.
- Strong mechanical connection.

Disadvantages:

- Requires correct tools to avoid damage.
- Slightly more expensive to manufacture.

Case-Back Screws

Rather than threading the entire back, multiple small screws along the perimeter secure the back plate to the case. It provides good sealing when combined with a gasket and is often seen in higher-end mechanical watches.

Advantages:

- Good sealing and rigidity.
- Service-friendly for watchmakers.

Disadvantage:

- Slightly slower to open/close (many screws).

Integrated/Monobloc Case Backs

The case and case back are one single piece of metal, especially in some vintage or specialized designs (e.g., one-piece middle case). In these designs, access to the movement is only from the front through the crystal or by removing other structural elements.

Advantages:

- Very secure construction.
- Good environmental protection.

Disadvantage:

- Very difficult to service.

Bayonet/Twist-Lock Case Backs

These use a twist-and-lock mechanism—similar to camera lenses—where the back is placed and twisted until it locks. They combine some advantages of snap and screw types.

Advantages:

- Easier to open than threaded backs.
- Better sealing than snap-ons.

Disadvantage:

- Less common; special tooling may be needed.

2.1.9 Watch Back Engravings

Wristwatches typically have special engravings on their case backs or case back. These engravings can be divided into serial engravings and individual engravings. The exact requirements for serial engravings can vary by country and watch type (e.g.,

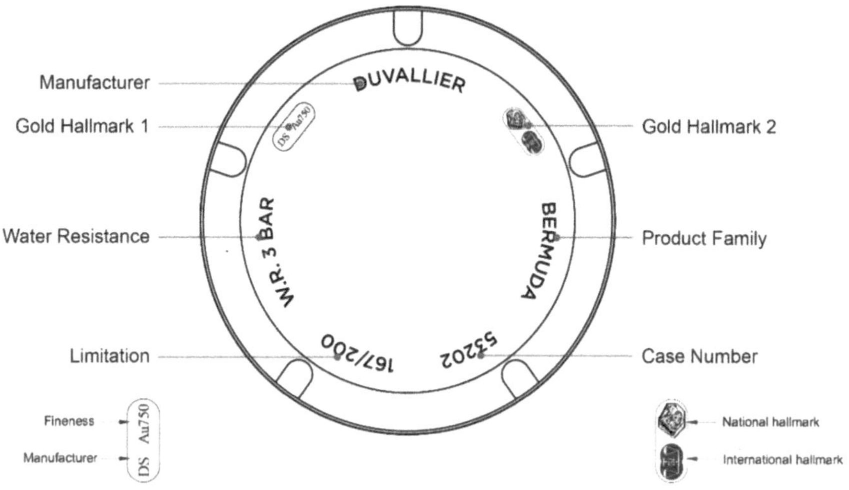

Fig. 2.10 Engraved watch back

consumer product vs. certified diving watch), but generally, the following information is either required by law or industry standard practice to be engraved or stamped on the case back of a wristwatch [49].

Material indications

If a watch case is made of precious metals (e.g., gold, silver, platinum), hallmarking regulations in many countries require that the purity (e.g., "18K", "750") be stamped. Figure 2.10 shows the base of a gold watch with the hallmarks required in Switzerland. Misleading inscriptions about materials (e.g., "stainless steel," "gold plated") are prohibited unless they are accurate.

Water resistance

If a watch is marketed as water-resistant, consumer protection and advertising laws require that the level of resistance (e.g., "100 m", "10 ATM", "30 BAR") be clearly indicated.

Manufacturer identification

The brand name must be present to ensure traceability of the product to its producer or distributor.

Serial or reference numbers

Often required for traceability, anti-counterfeiting, and warranty.

Name of the relevant product family

Not mandatory (e.g., pilot's watch).

Indication of a limited series

Not mandatory, customers often ask for a specific number (e.g., "8 out of 100").

Watch backs are engraved using the methods described below.

Hand Engraving

A traditional method where a skilled artisan uses small tools (burins or gravers) to carve designs or text directly into the metal surface.

- Unique, artistic, and highly detailed.
- Time-consuming and more expensive.

Machine Engraving (Rotary Engraving)

Uses a mechanical cutting tool guided by a machine to carve the design.

- More consistent than hand engraving, suitable for text and simple patterns.
- Limited in complexity compared to hand or laser engraving.

Laser Engraving

A precise, modern method where a laser beam etches the design onto the surface.

- Extremely accurate, allows for fine details (even photos or complex logos.
- Less "handmade" character can look very uniform.

Etching (Chemical Engraving)

Uses acid or chemicals to etch the design into the surface.

- Suitable for delicate or large surface designs, smooth finish.
- Less common today, it requires careful handling of chemicals.

CNC Engraving (Computer-Controlled Engraving)

A modern version of machine engraving where computer-controlled tools cut the design.

- High precision, repeatable, good for custom logos or mass production.
- Lacks the artisanal uniqueness of hand engraving.

In practice, laser engraving is now the most popular method for personalizing watch case backs, while hand engraving is still valued for luxury or artistic timepieces.

Hallmarks to identify the case material are an exception: they are always embossed.

2.1.10 *Gemstones*

Watches set with gemstones can be adorned using a variety of gem-setting techniques, each of which influences not only the visual appearance but also the durability, craftsmanship, and overall character of the timepiece [50].

Prong Setting

Small metal claws hold the gemstone in place. This technique allows maximum light to enter the stone, enhancing brilliance, and is often used for larger diamonds on bezels or dials, though the stones are slightly more exposed to impact.

Bezel Setting

A continuous metal rim surrounds the gemstone. This method offers excellent protection and a clean, modern esthetic, making it particularly suitable for sports or everyday watches. The bezel setting is valued for its robustness, although it can slightly reduce the stone's sparkle due to limited light exposure.

Pavé Setting

In this technique, numerous tiny stones are set closely together, each held in place by small metal beads, creating the illusion of a continuous, shimmering surface. Pavé is often applied to bezels, dials, or bracelets and is highly decorative, though it requires exceptional precision and maintenance to ensure long-term stability.

Channel Setting

Gemstones are placed in a row between two parallel metal walls without visible prongs. This method creates a sleek and orderly appearance and is frequently used for baguette-cut stones on bezels or bracelets. Channel setting combines elegance with relatively good protection, as the stones are secured along their edges.

Invisible Setting

This Method represents one of the most technically demanding techniques. Here, specially cut gemstones are locked together from underneath, with no visible metal holding them in place. The result is a seamless field of gemstones, offering a striking visual effect. However, this method requires extreme craftsmanship and is typically reserved for high-end jewelry watches due to its complexity and fragility.

Grain Setting or Bead Setting

This technique, closely related to pavé, involves fixing stones into small, drilled seats and securing them with tiny raised grains of metal. It is often used for delicate dial work and intricate decorative patterns, highlighting the artisanal aspect of watchmaking. Overall, the choice of setting method reflects a balance between aesthetics, wearability, and the intended use of the watch, showcasing the intersection of horology and fine jewelry.

When selecting gemstones for use in wristwatches, both the type of stone and the chosen cut are crucial, as watches are exposed to far more mechanical stress than most pieces of jewellery. Suitable gemstones must therefore combine beauty with high hardness, durability, and long-term stability, while the cut must support secure setting and visual harmony with the watch design.

Diamonds

Among all gemstones, diamonds are by far the most suitable and most used for watch settings. With a hardness of 10 on the Mohs scale, diamonds offer exceptional scratch resistance and durability, making them ideal for bezels, dials, and even bracelets. They can be used in a wide variety of cuts, including brilliant, baguette, princess, and rose cuts. Brilliant-cut diamonds maximise sparkle and are often used for pavé or prong settings, while baguette-cut diamonds are particularly well suited for channel and invisible settings due to their straight edges and uniform geometry.

Sapphires and Rubies

Sapphire and Ruby are stone varieties of corundum and are also highly suitable for watches, thanks to their hardness of 9 on the Mohs scale. Sapphires are available in a wide range of colors, allowing designers to create decorative or color-coded bezels and dials, while rubies add a strong, luxurious accent.

Emeralds

These stones can be used in watchmaking, but with greater caution. Their hardness is lower, and they often contain natural inclusions, making them more sensitive to shocks. For this reason, emeralds are typically reserved for low-impact areas such as decorative bezels or dress watches. The classic emerald cut, with its stepped facets, is preferred because it reduces internal stress within the stone and highlights color rather than brilliance.

In terms of cuts, brilliant cuts (see Fig. 2.11) emphasise light performance. They are ideal for decorative surfaces, while step cuts, such as baguette or emerald cuts, offer a clean, architectural look and are structurally well-suited to the linear design of watch bezels and bracelets.

The criterion "gem coverage" refers to the extent and distribution of gemstones across the watch's visible surfaces. It is one of the most important dimensions in classifying gem-set watches because it directly affects esthetics, craftsmanship complexity, value, and wearability.

This classification is typically divided into four principal categories (see Fig. 2.12).

Accented Gem-Set Watches

In this category, gemstones are used sparingly and serve primarily as subtle decorative highlights. Typical applications include diamond hour markers, small gemstone accents on the bezel, or decorative elements on the crown. The overall design of the watch remains dominant, with gemstones enhancing elegance rather than defining it. From a technical perspective, these watches require only limited intervention in the case structure and rely on standard setting techniques such as prong or light pavé.

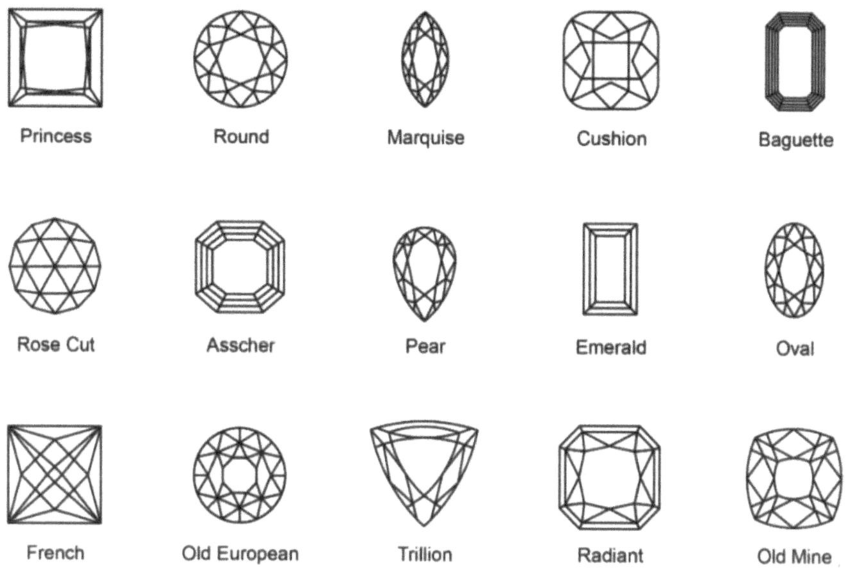

Fig. 2.11 Classification of gemstones cuts

Fig. 2.12 Classification of gemstones coverage

Partially Gem-Set Watches

These watches feature a more pronounced use of gemstones, with specific components of the watch being fully set while others remain unadorned. Common examples include bezels entirely set with diamonds, gem-set dials paired with plain cases, or bracelets incorporating gemstone-set links. This creates a deliberate contrast between polished metal and sparkling stones, resulting in a more distinctive visual identity. The craftsmanship involved is more demanding, as the integration of gem-set and non-set surfaces must be carefully balanced both aesthetically and structurally.

Fully Gem-Set Watches

In this category, nearly all visible surfaces—including the case, bezel, dial, and bracelet—are covered with gemstones. The underlying material of the watch, whether gold or steel, becomes visually secondary to the brilliance of the stones. These watches are immediately recognizable as high-luxury objects due to their intense light reflection and opulent appearance. Their production is extremely labor-intensive, requiring precise matching and cutting of stones, as well as advanced setting techniques such as pavé, baguette, or invisible setting.

High Jewelry (Haute Joaillerie) Watches

At the extreme end of the spectrum are high jewelry (haute joaillerie) watches, which represent the highest level of gem coverage and artistic ambition. In these pieces, the watch becomes first and foremost a work of jewelry, with timekeeping playing a secondary role. While they often feature complete or near-complete gem coverage, what distinguishes them is not merely the quantity of gemstones, but the artistic composition, the use of rare and exceptional stones, and the conceptual design. These watches frequently incorporate advanced and highly complex setting techniques such as invisible setting, and they are typically handcrafted in very limited numbers or as unique pieces. The integration of the movement into such elaborate jewelry structures requires exceptional technical expertise.

If a watch is intended to be offered in a version set with stones, this should be taken into account in the design of the standard version.

2.1.11 Crowns

Just as the rims give a car a special character, an interestingly designed crown can significantly improve the esthetics of a wristwatch [51].

Standard Crown

- The most common type, usually found on everyday watches.
- Simple push/pull mechanism to set the time and date.
- Not specifically designed for water resistance.

Screw-Down Crown

- Screws into the case to create a tight seal.
- Commonly used in dive watches or water-resistant timepieces.
- Provides extra protection against water, dust, and accidental adjustment.

Onion Crown

- Large, rounded, and textured, resembling the shape of an onion.
- Traditionally found on vintage pocket watches and some pilot watches.
- Easy to grip, even with gloves.

Diamond Crown

- Shaped like a cut diamond with multiple facets.
- A decorative variation that provides better grip and a stylish appearance.

Recessed or Hidden Crown

- Sits flush or partly hidden in the case.
- Gives the watch a sleek, modern look.
- Sometimes harder to operate, but it protects against accidental movement.

Left-Hand (Destro) Crown

- Placed on the left side of the case instead of the usual right.
- Designed for left-handed wearers or to avoid irritation on the wrist.
- Common in some professional tool watches.

Crowns with Guards

- Protected by metal shoulders on either side.
- Prevents accidental knocks and enhances durability.
- Often seen on military or sports watches.

Complicated Crowns (Multi-function)

- Used on watches with additional features (GMT, monopusher-chronograph, etc.).
- May have multiple positions for setting different complications.

Cabochon Crowns

- The word "cabochon" comes from the French *caboche*, meaning "head."
- Have a small, polished gemstone or decorative, rounded insert set into the tip.

The different crown shapes depicted in Fig. 2.13 can be distinguished:

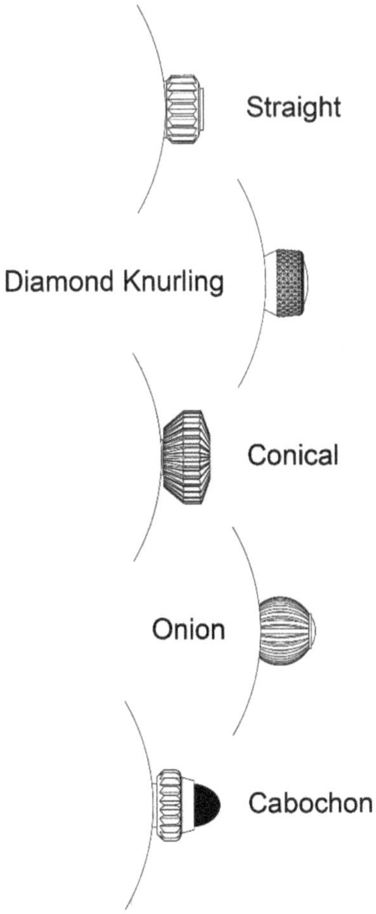

Fig. 2.13 Watch crown types

2.2 Dials

The dial is the most visually striking element of any watch; it is its face. It significantly contributes to giving a watch a particular feel. A clear design with easily readable numbers emphasizes the functional style, while a complex surface and fine indices emphasize the luxurious character of a watch [52].

2.2.1 Elements of a Dial

A wristwatch dial can include a wide range of elements, depending on the watch's function, style, and level of complexity. The most common and optional elements are shown below in Fig. 2.14.

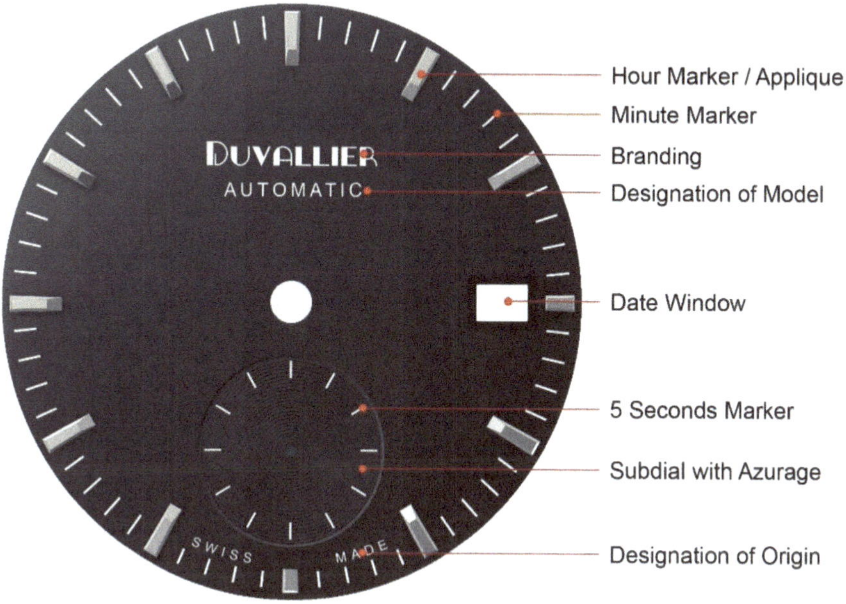

Fig. 2.14 Dial with labels for the various elements

The *azurage* of the sub dial shown above is a form of fine circular or spiral texturing applied to the surface of a dial. It is created by manually or mechanically engraving very fine, concentric lines into the metal surface using a rotating tool. An azurage is commonly found on sub-dials of chronographs, small seconds displays, or power-reserve indicators, where it provides a subtle contrast to smoother dial surfaces such as sunburst, lacquered, or matte finishes.

The dials of some high-quality watches feature metal rings that are positioned over the central opening for the hands or serve as visual boundaries for sub-dials. In watchmaking terminology, these rings, also made of precious metals, are called *canons*.

Dials typically have 2–4 cylindrical dial feet on their underside. Their position corresponds to the holes provided for them in the watch movement.

2.2.2 *Surface Treatment*

The surface of a watch dial plays a crucial role in both esthetics and functionality. Beyond color, the texture and finish influence how light is reflected, how legible the watch is, and how luxurious or sporty it appears.

The dial surface is influenced not only by the material used, but also by its machining. Common methods include [53].

Mechanical Surface Treatment

- Grinding
- Polishing
- Brushing
- Blasting

Chemical Surface Treatment

- Electroplating
- Anodizing

Coatings

- Painting
- Powder Coating

The following surface types are distinguished.

Matte (Flat) Dials

A matte dial (see Fig. 2.15) has a non-reflective, flat surface.

- Characteristics: Minimal glare, understated appearance
- Effect: Excellent legibility, tool-watch or minimalist feel
- Typical use: Pilot watches, military watches, Bauhaus-inspired designs

Fig. 2.15 Dial with matte finish

Fig. 2.16 Dial with sunburst finish

Sunburst (Soleil) Finish

A sunburst dial (see Fig. 2.16) features fine radial brushing from the center outward.

- Characteristics: Dynamic light reflections depending on angle
- Effect: Elegant, lively, and visually engaging
- Typical use: Dress watches, luxury sports watches

Glossy/Lacquered Dials

A glossy dial (see Fig. 2.17) is polished and often coated with multiple layers of lacquer.

- Characteristics: Smooth, mirror-like surface
- Effect: High contrast, refined and luxurious look
- Typical use: Formal dress watches, enamel-style designs

Grainy/Textured (Frosted, Sandblasted) Dials

A dial with a surface (see Fig. 2.18) has a fine, irregular texture.

- Characteristics: Subtle roughness, diffused reflections
- Effect: Technical, contemporary, or artisanal appearance
- Typical use: Tool watches, independent or avant-garde brands

Fig. 2.17 Dial with glossy finish

Fig. 2.18 Dial with frosted finish

Guilloché Dials

Guilloché refers to intricate, engraved geometric patterns, traditionally done by hand or engine-turning. Figure 2.19 shows a dial with a guilloché finish.

- Characteristics: Repeating decorative motifs (e.g., Clous de Paris)

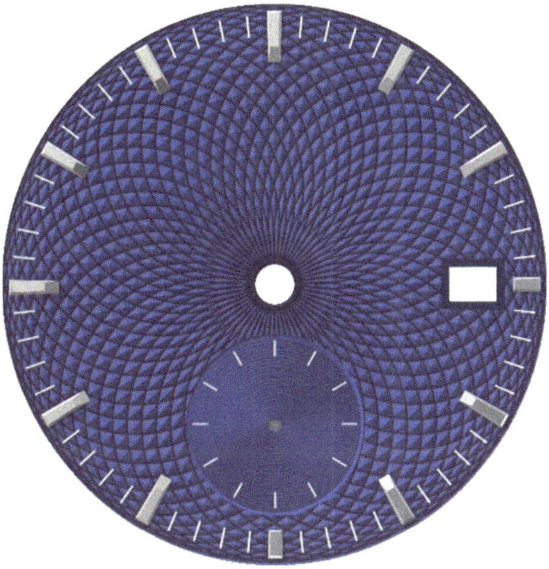

Fig. 2.19 Dial with guilloché finish

Fig. 2.20 Dial with Grand Feu finish

- Effect: Strong sense of craftsmanship and tradition
- Typical use: High-end dress watches, classical complications

Fig. 2.21 Dial with Fumé finish

Enamel Dials (Grand Feu, Cloisonné)

An enamel dial (see Fig. 2.20) is made by firing powdered glass onto a metal base at high temperatures.

- Characteristics: Perfectly smooth surface, deep glossy color
- Effect: Exceptional depth, longevity, and prestige
- Typical use: Luxury and collectible watches

Fumé/Gradient Dials

Such a dial (see Fig. 2.21) transitions from a lighter center to darker edges.

- Characteristics: Gradual color gradient
- Effect: Depth, drama, and visual focus toward the center
- Typical use: Modern luxury watches, vintage-inspired designs

Patterned or Fabric-Inspired Dials

Examples of patterned dials (see Fig. 2.22) include waffle patterns, linen textures, or tapisserie motifs.

- Characteristics: Repeating structured patterns
- Effect: Sporty or technical identity, strong brand recognition
- Typical use: Sports watches (e.g., integrated-bracelet designs)

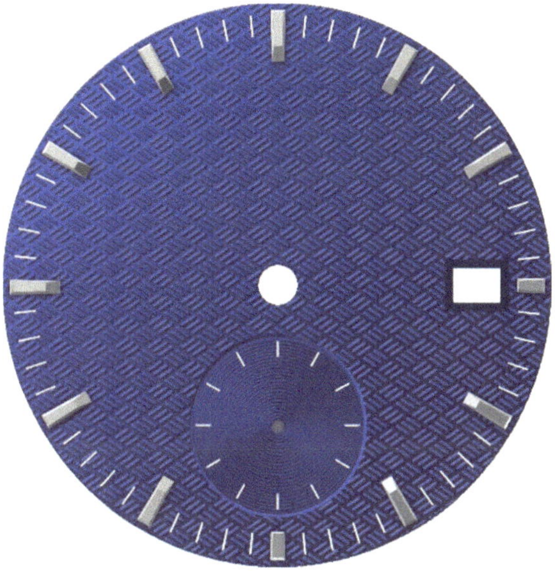

Fig. 2.22 Dial with patterned finish

2.2.3 Print

The dials of wristwatches can be printed and decorated using several different techniques. The choice of method affects precision, durability, cost, and visual appearance. The two most important printing and marking methods are explained below:

Pad Printing (Tampography)

Pad printing is the most common technique used for watch dials. A soft silicone pad transfers ink from an engraved plate onto the dial surface. This method allows very fine details, sharp edges, and accurate positioning, even on curved or textured dials. It is typically used for logos, minute tracks, numerals, and small text.

Screen Printing (Silk-Screen Printing)

In screen printing, ink is pressed through a fine mesh stencil onto the dial. This technique is suitable for larger areas of color and bold graphics. It is less precise than pad printing but allows thicker ink layers, which can improve color intensity and durability.

2.2.4 Proportions

In watchmaking, *proportion* isn't just about how big or small a dial is—it's about how the elements on the dial relate to each other visually and functionally. Good proportions help make a dial:

- easy to read at a glance
- visually balanced
- pleasant to look at
- harmonious with the watch case and hands

A well-proportioned dial, as seen in Fig. 2.23, typically occupies most of the case front—but not overly so. Designers often aim for around 80–95% of the wrist width, matched to the wearer's ergonomics and visual balance.

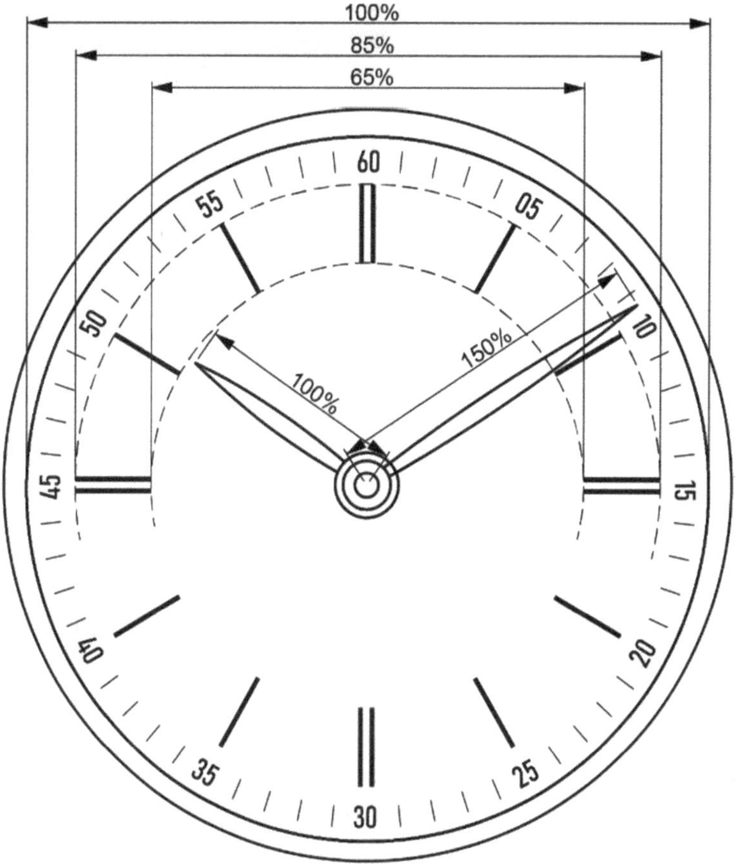

Fig. 2.23 Popular proportions of a watch dial including hands

The *effective dial area* (not just absolute size) should feel balanced with the bezel and lugs. A bezel that is too thick can make a dial look smaller and emptier.

Most traditional dials use symmetrical layouts (e.g., twelve equidistant markers) because symmetry is inherently pleasing and makes time easier to read [54].

On rectangular wristwatches, designers deliberately use the *Golden Ratio* (~1:1.618) in dial layouts to improve esthetic harmony. (A famous example is the Jaeger-LeCoultre Reverso.) Typical applications include:

- spacing between the main dial and the minute track
- ratios of dial elements and sub dials
- size relationships among hands and markers

2.2.5 Colors

The dial color of a watch isn't just visual decoration—it communicates meaning, conveys personality, and can influence how the wearer and observers perceive the watch and the brand.

The colors of a dial can be distinguished from the following aspects:

Color Mixing:

- Primary colors
- Secondary colors
- Tertiary colors

Subjective Temperature:

- Warm
- Cool

Tonal Value:

- Light
- Dark

Red, blue, and yellow are known as primary colors (1). They cannot be created by mixing other colours. The secondary colors (2)—green, orange, and violet—are created by mixing two primary colors. Tertiary colors (3) occur when a primary color is mixed with a secondary color.

Figure 2.24 shows the model of a color circle used by Isaac Newton, among others, to explain the relationships between the hues.

Monochromatic colors differ only in their brightness, i.e., in their proportion of black and white. These tone-on-tone combinations are very harmonious and often appear elegant. Complementary colors are colours that are directly opposite each other on the color wheel. They create a strong contrast and are very striking when used side by side. Analogous colors are those next to each other on the color wheel. These combinations are generally harmonious and create a calming effect. Triadic

Fig. 2.24 Color wheel

colors are evenly distributed on the color wheel. These combinations offer a vibrant yet balanced color harmony. Tetradic colors consist of two pairs of complementary colors. This combination creates a rich, dynamic color palette. Split complementary colors are combined from one color and the two colors adjacent to that color's complement. They offer contrast but are less intense than pure complementary colors. Square colors are evenly distributed on the color wheel. This combination is balanced yet vibrant. Figure 2.25 shows the described color constellations.

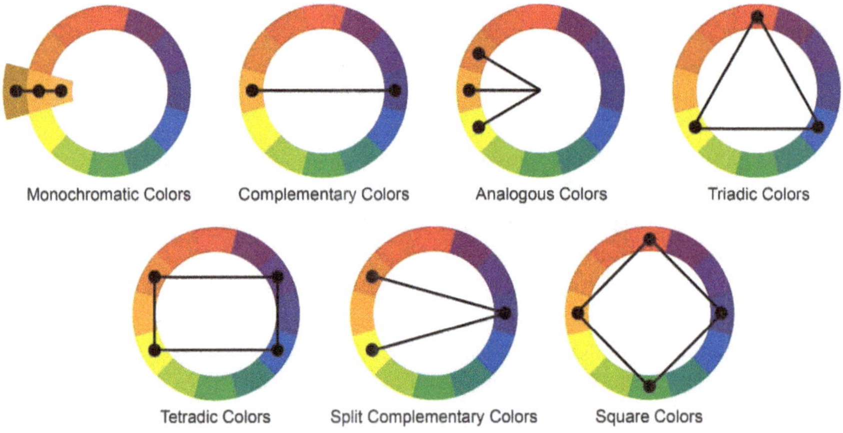

Fig. 2.25 Color constellations

Designers must understand what colours *say* to people. Colors trigger cultural and emotional associations that the wearer notices even unconsciously [55].

- Red is bold, energetic, and attention-grabbing.
- Green is linked to nature, renewal, and calm—and can feel both fashionable and unique.
- Blue suggests stability and elegance, often associated with both sport and classic watches.
- Black evokes sophistication, seriousness, and timelessness.
- White/silver tends to signal clarity, minimalism, and traditional elegance.

High contrast between dial color and hands/indices improves legibility. For instance, a light dial with dark hands or vice versa.

Neutral tones (black, white, grey) remain timeless classics.

Fashionable colors reflect contemporary trends and help the watch feel relevant yet personal.

Designers must consider *where and how* the watch will be worn—a flamboyant red dial may suit casual attire but feel out of place in a conservative setting [56].

2.2.6 Typography

The numerals (and any text) on a watch dial are not just functional elements—they are core design features that give a watch its personality and communicate subtle stylistic cues to the viewer. Typography includes all written elements on the dial: numbers, indexes, brand logos, and letterforms. This is one of the key elements that differentiates one watch design from another [57].

Different numeral types—such as Arabic numerals, Roman numerals, or stylized numerals—immediately convey different stylistic associations:

- Roman numerals often feel classical, traditional, or dressy.
- Arabic numerals tend to appear more utilitarian or casual.
- Stylized numbers (e.g., Art Deco-inspired or highly modified forms) can evoke vintage, modern, or avant-garde themes.

All numerals can be displayed in different fonts, which can be divided into the following groups:

Serif

Serif fonts have small lines or strokes (serifs) attached to the ends of letters. They are seen as traditional, elegant, and highly readable in print and are often associated with authority, history, and professionalism.

Sans-Serif

Sans-serif fonts lack the extra strokes of serif fonts, giving them a clean, modern, and minimal appearance. They are straightforward, easy to read on screens, and often convey clarity and simplicity.

Script

Script fonts mimic handwriting or calligraphy, with flowing, connected strokes. They feel personal, artistic, and often elegant or romantic, though sometimes less legible in long texts.

Decorative

Decorative (or Display-) fonts are designed primarily for visual impact rather than readability in body text. They are expressive, playful, and highly distinctive, often used for headlines, posters, or branding.

Monospace

Monospace fonts give each character the same fixed width, which makes them look mechanical and structured. They are practical, technical, and often used for coding or tabular data where alignment is important.

Blackletter

Blackletter (or Gothic) fonts are highly ornamental, with dense, angular strokes that resemble medieval manuscripts. They convey tradition, formality, or historical gravitas, but can be hard to read in longer texts.

Figure 2.26 shows numbers in exemplary types from the different groups of fonts.

Some manufacturers go beyond off-the-shelf fonts and create bespoke numeric typefaces. This contributes to a unique character. Some brands choose specific typefaces corresponding to their company's corporate design.

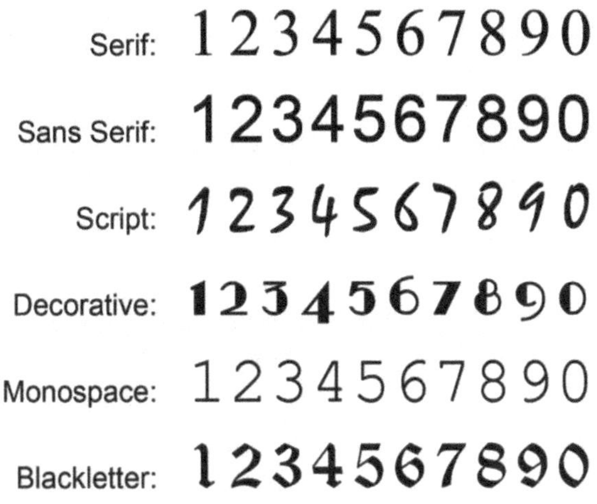

Fig. 2.26 Numerals in different fonts

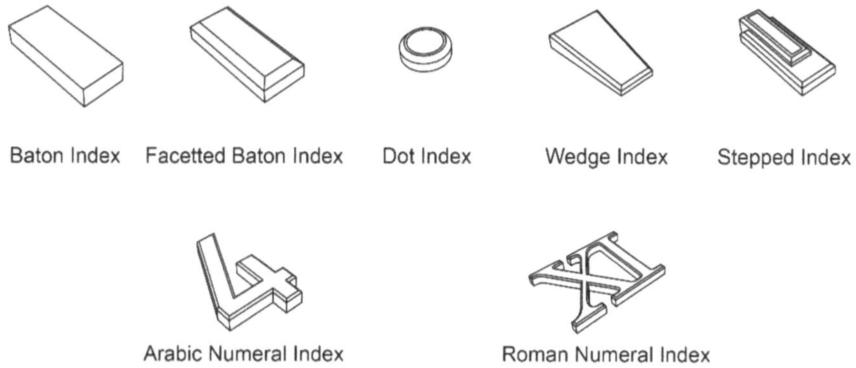

Fig. 2.27 Different types of appliques

Typography on a watch isn't solely about esthetics—it influences functionality and visual balance. Well-designed numerals improve legibility at small scales, which is critical for quickly telling time. The proportions, spacing, stroke weight, or ornamentation of numerals can make a watch feel precise and technical or soft and elegant.

Users often associate certain numeral styles with cultural or historical references:

- Classic serif numerals may feel luxurious or historic.
- Clean sans-serif figures can suggest modernity and minimalism.

Appliques

Metal indices (also called applied indices or appliques, see Fig. 2.27) on watch dials come in several different forms, each serving both a functional and an esthetic purpose [58].

Baton Index

- Consists of simple, straight metal bars.
- Used on modern and minimalist watches.

Facetted Baton Index

- Features angled or multi-faceted surfaces that reflect light.
- Adding depth and visual interest to the dial.

Dot Index

- Circular metal markers that may be flat or slightly domed.

Wedge Index

- At the outer edge of the dial and taper inward.

Stepped Index

- Consists of several stacked or raised surfaces.

Arabic Numeral Index

- Metal numerals as hour markers.
- Give the watch a more traditional or instrument-like character.

Roman Numeral Index

- Commonly found on dress watches.
- Emphasize elegance and classic design.

In addition to shape, metal indices differ in finishing and construction. They may be polished, brushed, or sandblasted, and are often rhodium-plated, gold-plated, or black-coated to match the overall design of the watch. Many applied metal indices also incorporate luminous material, either as an inlay, a filled center, or a coated surface, ensuring readability in low-light conditions. Together, these variations in shape, finish, and execution allow metal indices to significantly influence the character, legibility, and perceived quality of a watch dial.

2.2.7 *Luminous Material*

Luminous materials used on the hands, dials, and indices of wristwatches are designed to ensure readability in low-light or dark conditions, and they differ significantly in composition, performance, and historical context [59].

Radium

One of the earliest types of luminous material was radium-based lume, used mainly in the early to mid-20th century. Radium paint glowed continuously without needing prior exposure to light because it was radioactive. However, due to its severe health risks, it was discontinued and is no longer used in modern watches.

Tritium

Tritium is a radioactive hydrogen isotope that emits a soft, constant glow and was widely used from the 1960s through the 1990s. It is much safer than radium but still mildly radioactive, which led to regulations on its use. Over time, tritium lume fades and eventually stops glowing. Some watches use tritium gas tubes, which are tiny, sealed glass tubes filled with tritium gas and coated internally with phosphor. These tubes glow continuously for many years without charging from light and are especially popular in military and tactical watches.

Super-LumiNova®

The most widely used luminous material in modern watchmaking is photoluminescent, non-radioactive lume, commonly known under brand names such as Super-LumiNova®. This type of lume absorbs light from natural or artificial sources and

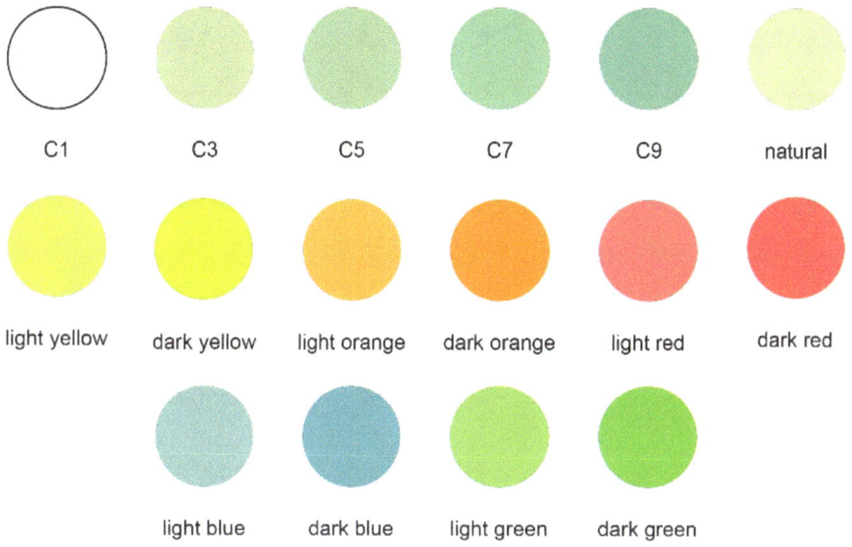

Fig. 2.28 Most common Super-LumiNova® colors

then emits it in darkness. Its brightness depends on the amount of light absorbed and gradually fades over time, but it can be recharged repeatedly without degradation. Photoluminescent lume is available in various colors in daylight, such as white, cream, or pale green, while glowing in green, blue, or other hues in the dark (see Fig. 2.28). It is safe, stable, and highly versatile, making it the industry standard today.

In addition to chemical composition, luminous materials differ in how they are applied. On hands and indices, lume may be painted on, filled into recesses, or applied as solid luminous blocks or inlays, which can enhance brightness and longevity. On dials, lume may be printed directly onto the surface, applied to raised indices, or placed beneath translucent materials to create a softer glow. The thickness and quality of application play a crucial role in luminosity performance.

2.2.8 Layout

In the context of wristwatches, a dial layout refers to the overall arrangement and design of all visual elements on the watch dial that help display the time and other information.

This includes, most importantly, the placement and style of hour indications, such as numerals or simple hour markers. A dial layout defines which hours are numbered, where those numerals appear, and how they are combined with markers.

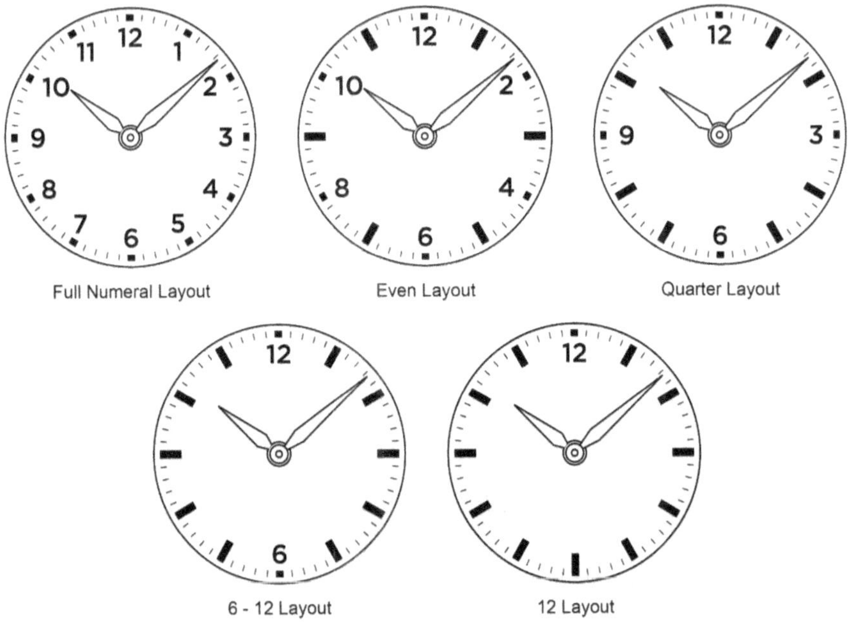

Fig. 2.29 Layout-types of wristwatch dials

Numerals do not always appear at all twelve-hour positions; instead, designers often combine numerals with hour markers for esthetic and functional reasons. The layouts range from fully numbered dials to highly abstract designs with minimal numerals.

Figure 2.29 shows the main layout types.

Beyond hour indications, the dial layout also encompasses the position of the hands, sub-dials (for chronographs or calendars), date windows, logos, and other inscriptions. The layout determines how these elements interact visually and functionally, ensuring legibility while supporting the intended esthetic.

In summary, a watch dial layout is the *conceptual structure* of the dial, balancing readability, functionality, and design language. It plays a key role in defining the character of a watch, whether it is classical, sporty, minimalist, or highly technical [60].

The so-called *California dial* (see Fig. 2.30) is a distinctive type of watch dial that combines both Roman and Arabic numerals on the same face. Traditionally, the upper half of the dial (often from about 10 o'clock to 2 o'clock) shows Roman numerals, while the lower half (roughly 4 o'clock to 8 o'clock) displays Arabic numerals. The remaining hour positions (usually 3, 6, and 9) are often indicated by simple markers or dashes, and the 12 o'clock position may be marked with a triangle or inverted arrow. This layout creates a visually interesting contrast and was originally designed to enhance legibility—making it easier to read the time quickly at a glance, particularly in low-light conditions or during activity [61].

Fig. 2.30 California-dial

2.2.9 *Chronograph Scales*

A chronograph is a type of watch that includes a stopwatch function. Many chronograph watches have scales around the dial to calculate specific measurements based on time. These include tachymeter, telemeter, and pulsometer [62].

Tachymeter

The purpose of a tachymeter is to measure speed over a known fixed distance (1 kilometre or 1 mile).

When creating a tachymeter scale (see Fig. 2.31), the angle for the speed (e.g., 94) ranges is calculated as follows:

$$\alpha = \frac{60 \times 360}{\text{kmh/mph}}$$

How it works

- Start the chronograph when the object (e.g., car) passes the start point.
- Stop it when the object passes the end point (after 1 km or 1 mile).
- Read the speed (in km/h or mph) directly on the tachymeter scale where the chronograph second hand stops.

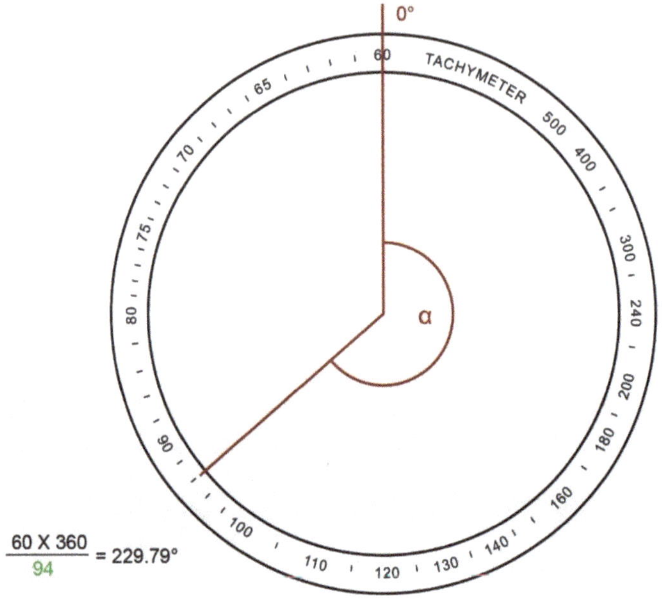

Fig. 2.31 Tachymeter scale of a chronograph

Telemeter

The purpose of a telemeter is to measure the distance to an event that is both visible and audible (like lightning and thunder). The telemeter scale is based on the well-known physical fact that light travels faster than sound. The formula is essentially simple:

$$\textbf{Distance} = \textbf{Time Difference} \times \textbf{Speed of Sound}$$

The telemeter scale (see Fig. 2.32) is usually given in kilometres and assumes a speed of sound of approximately 343 m/s (at 20 °C).

When creating a telemeter scale, the angle for the distance (e.g., 12 km) is calculated as follows:

How it works

- Start the chronograph when you see the event (e.g., lightning).
- Stop it when you hear the sound (e.g., thunder).
- The telemeter scale gives you the distance to the event (usually in kilometres or miles), based on the speed of sound (~343 m/s in air).

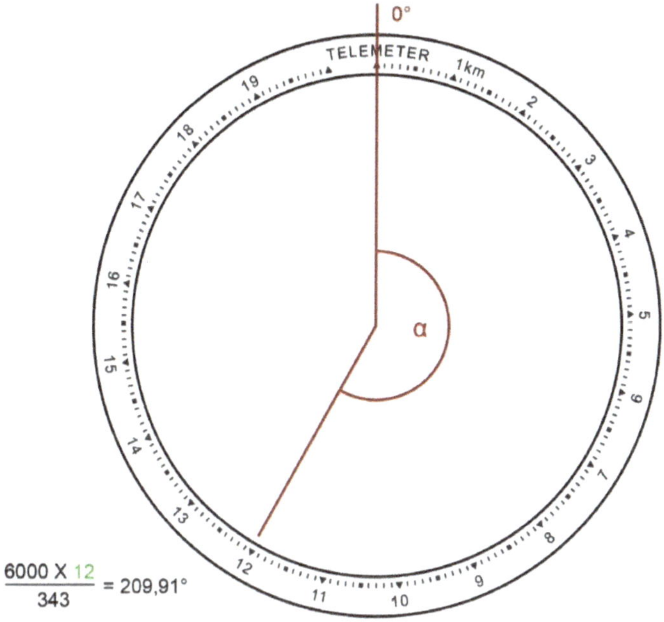

Fig. 2.32 Telemeter scale of a chronograph

Pulsometer

The purpose of a pulsometer is to measure the heart rate (pulse). Since the pulsometer scale is time-dependent (like the tachymeter scale), it is not linear, but follows a reciprocal relationship between beats per minute (bpm) and time, and thus also a logarithmic arrangement on the circle.

The angular position of a bpm value on the dial is determined by the corresponding time.

When creating a speedometer scale like shown in Fig. 2.33, the angle for the speed (e.g., 50 bpm) range is calculated as follows:

$$\alpha = \frac{10800}{\text{bpm}}$$

How it works

- Start the chronograph and count a set number of heartbeats (often 15 or 30).
- Stop it after counting the beats.
- The pulsometer scale gives the heart rate in beats per minute (bpm).

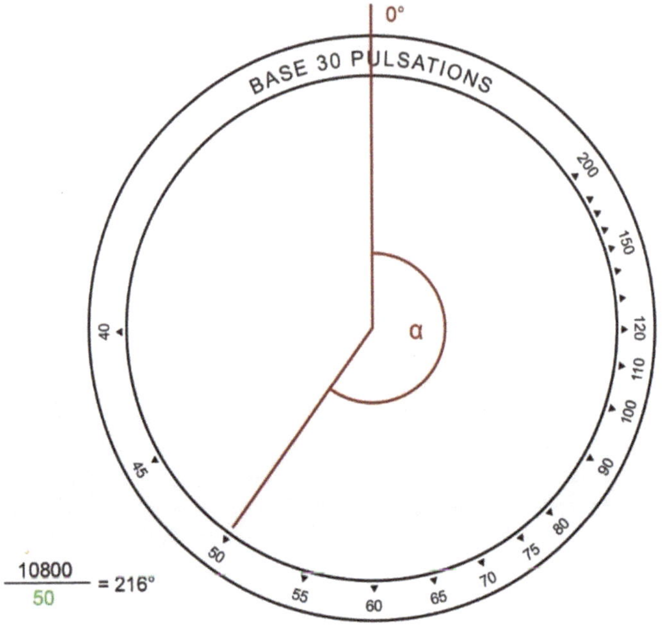

Fig. 2.33 Pulsometer scale of a chronograph

2.2.10 *Scales for Special Applications*

Watches designed primarily for specific applications may feature unusual scales on their dials and bezels. Some examples are shown below.

Regatta Countdown

A regatta watch, also known as a sailing regatta timer or sailing start watch, is a specialized timekeeping device designed for sailing regattas. The main function of a regatta watch is the countdown timer (see Fig. 2.34), which allows sailors to accurately track the remaining time until the start of the regatta. The timers can be set to the predetermined start time of the regatta and then count down the remaining minutes and seconds until the start.

How it works

- As soon as the regatta organizer's visual and/or audible signal is seen or heard 5 min before the start of the race, the countdown on the clock should be started.
- The five circular windows on the clock face change color successively to indicate the elapsed time until the actual start.
- Taking into account the displayed time remaining until the start, the sailor has the opportunity to manoeuvre his boat into the ideal position on the starting line.

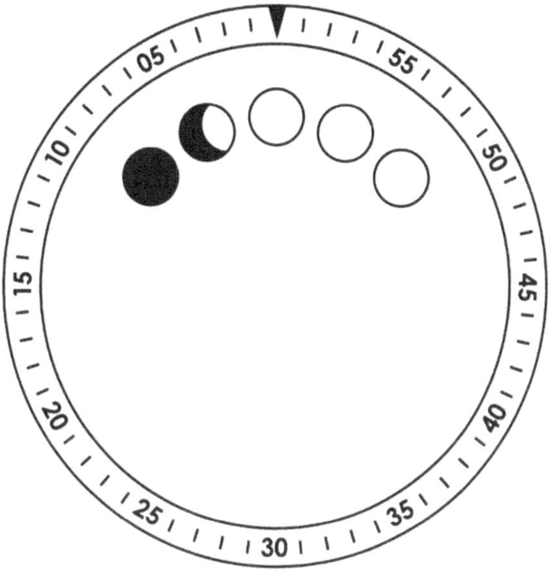

Fig. 2.34 Countdown scale of a Regatta watch

Calculator

A calculator bezel (see Fig. 2.35) functions as a circular slide rule, turning the watch into a compact analogue calculating instrument. Its operation is based on two logarithmic scales: the outer scale on the rotating bezel and the inner fixed scale around the dial. Because these scales are logarithmic, the physical distance between numbers represents a ratio rather than a simple linear value. By rotating the bezel, these distances can be added or subtracted, which corresponds mathematically to multiplication and division.

When creating a calculator scale, the angle of certain values (e.g., 4) is calculated as follows:

$$\alpha = 360 \cdot \log_{10}(X)$$

How it works

- **Multiplication**

  Example: **12 x 3**

Steps

1. Rotate the bezel so that **10 on the outer scale** aligns with **12 on the inner scale**
2. Find **30 on the inner scale** (= 3.0)
3. Read the corresponding value on the outer scale → **36**

   Result: **36**

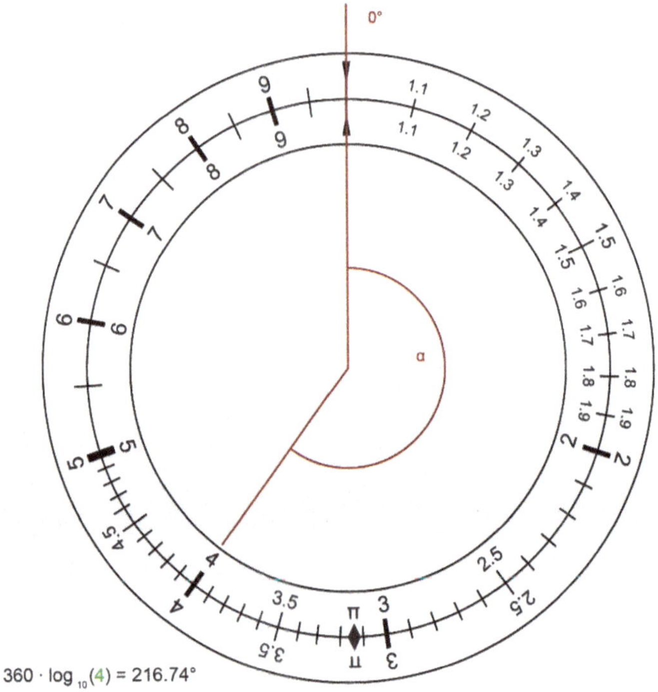

Fig. 2.35 Circular slide rule calculator

- **Division**

  Example: **36 ÷ 3**

Steps

1. Align **30 on the outer scale** with **36 on the inner scale**
2. Look at **10 on the outer scale**
3. Read opposite on the inner scale → **12**

  Result: **12**

- **Percentage calculations**

  Example: **15% tip on 80**

1. Align **10 outer** with **80 inner**
2. Find **15 outer**
3. Read opposite on inner → **12**

  Result: 15% of 80 = **12**

- **Speed**

  If 120 km are covered in 1.5 h:

1. Align **15 outer** with **12 inner**
2. Read **10 outer**
3. Opposite inner ≈ **80**

Result: **80 km/h**

One of the most practical uses of a calculator bezel is unit conversion. Because the scales express proportional relationships, fixed conversion factors such as kilograms to pounds or kilometres to miles can easily be set. If 22 on the bezel is aligned with 10 on the inner scale, the watch effectively represents the conversion factor of 1 kg to 2.2 pounds. From there, values such as 50 kilograms can quickly be converted to approximately 110 pounds.

The slide-rule bezel is equally useful for speed, travel, and fuel calculations, which reflects its roots in aviation and engineering instruments. Distances, times, average speeds, consumption rates, and estimated arrival times can all be determined by setting known values against each other and reading the proportional result from the opposite scale.

Naturally, the system is designed for quick approximations rather than exact precision. Because of the compact dimensions of the bezel, results are normally accurate to about two or three significant digits, which is nevertheless more than sufficient for engineering estimates, travel calculations, or rapid everyday conversions.

No-Decompression

Before the development of dive computers, watches with special scales helped divers to identify the maximum dive time at a given depth, after which they could surface without decompression stops. An example, the Fortis Marinemaster, is shown in Fig. 2.36.

How it works

- Using a second crown (typically at 4 o'clock), you rotate the internal ring until your planned maximum depth (in feet) lines up with the marker triangle at 12 o'clock. Depth values are printed around the outer inner scale (e.g., 50, 60, 80, etc.).
- You find your depth (Column **M** = depth in meters, Column **F** = depth in feet) and read the maximum bottom time (in minutes) before mandatory decompression stops are required.
- You then use the normal minute hand (or bezel reference) to track elapsed dive time.

Camera Exposure Settings

The bezel shown in Fig. 2.37 (System Horage Lensman) is designed as an analogue calculation tool for determining correct camera exposure settings, which makes it especially useful for photographers working with manual or film cameras. The rotating bezel aligns ISO values with lighting-condition symbols so that the appropriate shutter speed and aperture combinations can be read directly from the dial.

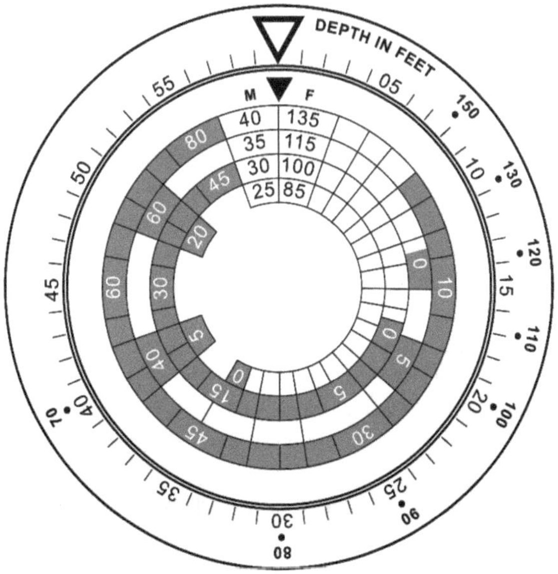

Fig. 2.36 Circular no-decompression scale of a diver's watch (courtesy of Fortis)

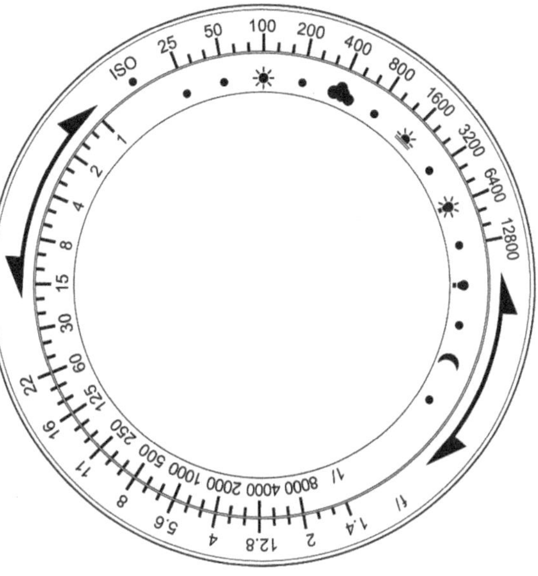

Fig. 2.37 Circular aperture and shutter time determination scale (courtesy of Horage)

The benefit of this bezel lies in its speed and independence from electronics.

How it works

- First, determine the ISO setting of your film or camera sensor, for example, ISO 200. Next, assess the ambient light—for instance, bright sunlight, cloudy weather, or indoor light.
- Then rotate the bezel until the chosen ISO value lines up with the corresponding light symbol.
- Once this is done, read the resulting values on the inner scale: you will find combinations such as 1/250 s at f/8 or 1/125 s at f/11. These are equivalent exposure settings, meaning you may choose the one that best suits your photographic intention. For example, a faster shutter speed helps freeze motion, while a wider aperture creates shallower depth of field.

2.2.11 Hands

A characteristic feature of an analogue time display is the use of hands. An analogue display emphasizes the *cyclical nature of time*, making it easy to see relationships such as how much time has passed or how much remains until the next hour [63].

Hour and Minute hands

The following points should be considered when designing hour and minute hands [64, 65].

Legibility

The hands must be clearly distinguishable from each other and from the dial under all lighting conditions. This includes choosing appropriate lengths, widths, and shapes so that the hour, minute, and second hands can be easily read at a glance. Adequate contrast in color or finish between the hands and the dial is particularly important.

Proportions and Geometry

The hour hand should be shorter and usually broader, the minute hand longer and slimmer, and the second hand clearly identifiable, often by a thin profile or a distinctive tip. Proper alignment with the dial's markers ensures accurate and intuitive time reading. The length ratio between the hour hand and the minute hand is usually 1 to 1.5. The minute hand should reach close to the minute track or index for accurate reading.

Material and Finish

Watch hands should be lightweight to minimize energy consumption and wear on the movement, yet rigid enough to resist bending. Common materials include steel, brass, titanium, gold, or precious alloys, depending on the watch's positioning and price point. Finishes such as polishing, brushing, or coating (e.g., bluing or PVD) affect both appearance and readability by controlling reflections.

Bluing watch hands is a traditional heat-treatment process used in watchmaking to create the deep blue color often seen on high-quality steel hands and screws.

Besides its decorative appearance, the process also offers a small degree of corrosion resistance.

PVD coating of watch hands is a modern surface-finishing process used in watchmaking to improve durability, corrosion resistance, and appearance. "PVD" stands for *Physical Vapor Deposition*. The method creates an extremely thin but hard coating on the metal surface of the hands.

Visibility in Low-Light Conditions

Many watch hands incorporate luminous materials, allowing the time to be read in darkness. The size, placement, and color of the luminous elements should be balanced so they enhance readability without overwhelming the design.

Mechanical Compatibility

Designing hands also means considering how they interact with the watch movement: Hands must be balanced so they don't strain the movement or cause friction. They must be mounted precisely to avoid hitting each other or the crystal/dial.

Design Coherence and Brand Identity

The style of the hands should match the overall character of the watch—whether classic, sporty, minimalist, or avant-garde—and reflect the brand's visual language. Well-designed hands not only tell time accurately but also contribute significantly to the watch's emotional appeal and identity.

Figure 2.38 shows the most common shapes for pairs of hour and minute hands.

Central Second-hand and chronograph's stop hands

When designing a central second-hand or a chronograph's stop hand (see Fig. 2.39), legibility is essential. The hand must be easy to distinguish from other hands, often through differences in length, thickness, color, or shape. It typically extends all the way to the outer minute or seconds track to allow precise reading of elapsed time. A clearly defined tip—sometimes needle-shaped, sometimes with a fine pointer or contrasting accent—helps the wearer align the hand accurately with the scale.

Weight and balance are critical technical considerations. Because the chronograph's stop hand is driven by a complex mechanism and may start, stop, and reset frequently, its mass must be carefully controlled. An unbalanced or overly heavy hand can negatively affect amplitude, accuracy, and long-term wear of the movement. For this reason, designers work closely with engineers to ensure proper counterweighting and smooth operation.

Sub Dial Hands

While creating small sub-dial hands (often called counter or register hands) for wristwatches, special attention must be paid to clarity, proportion, and mechanical constraints. Although these hands are smaller and less prominent than the central hands, they play an important role in readability and overall visual harmony.

Sub dial hands (see Fig. 2.40) must be clearly distinguishable from the dial surface and from each other. Their length should be precisely matched to the scale of the

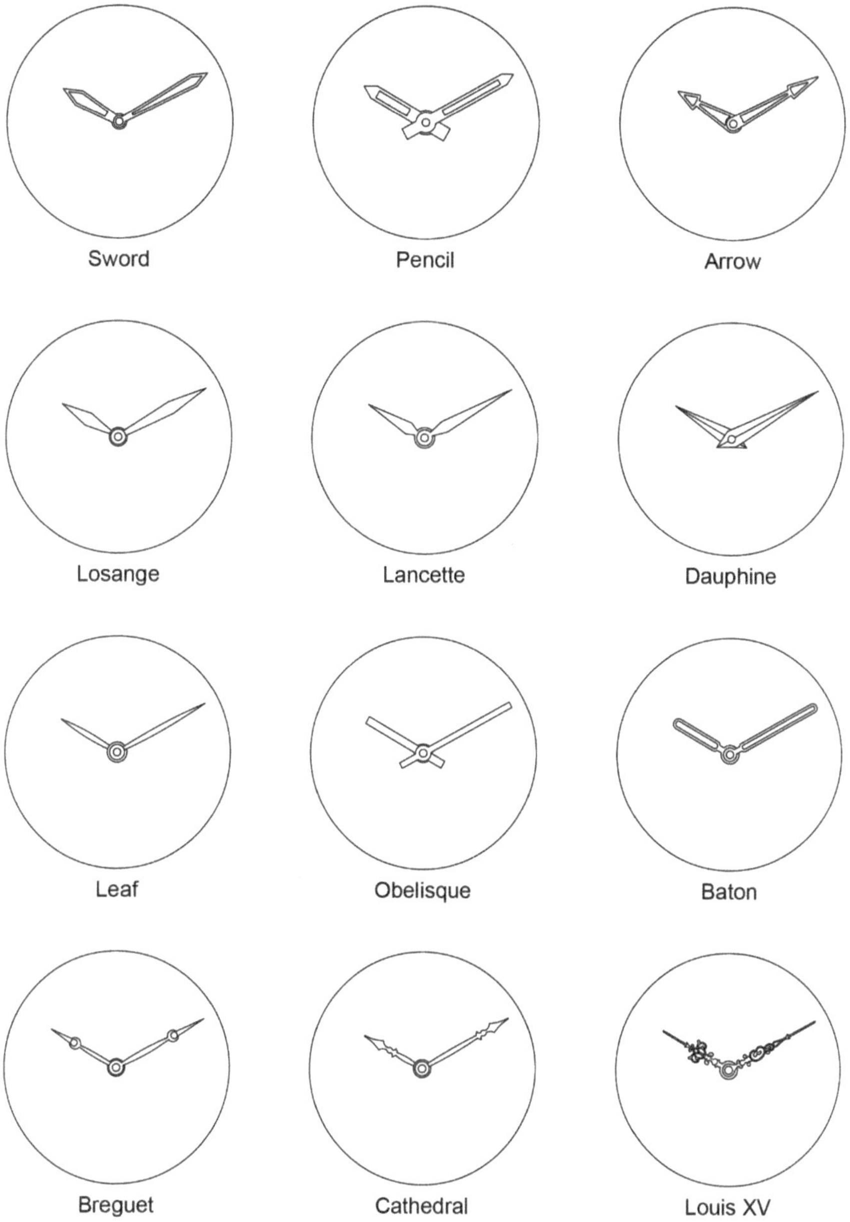

Fig. 2.38 Traditional watch hand shapes

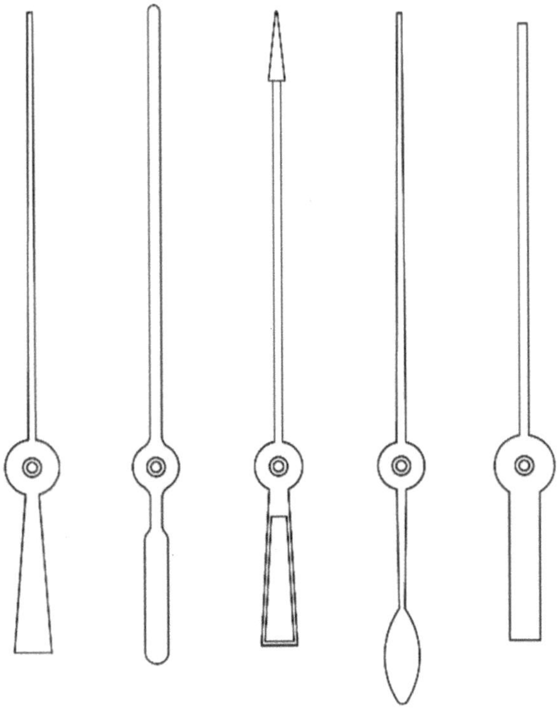

Fig. 2.39 Popular stop hand shapes

sub-dial: long enough to reach the markings for accurate reading, but short enough to avoid visual clutter or overlap with sub-dial borders. Fine tips and clean outlines help ensure precision despite the reduced scale.

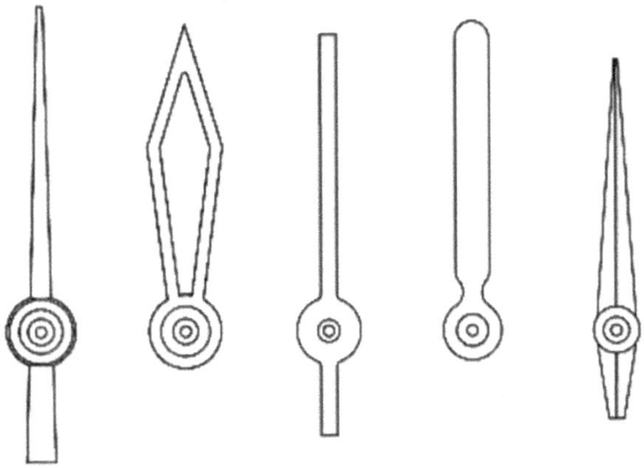

Fig. 2.40 Popular small hand shapes

2.3 Straps and Bracelets

The following statement regarding esthetics can be made for the vast majority of watch straps: the width should be approximately 50% of the case diameter to ensure mechanical and visual balance. Strap material significantly influences perceived size and mass: metal bracelets increase apparent volume and weight, while leather, fabric, or silicone straps produce a slimmer visual profile. Figure 2.41 shows the usual designations for bracelets of different lengths.

2.3.1 Leather Straps

Leather watch straps have distinct advantages over other types of watch bands, but they also come with certain limitations. One of their greatest strengths is comfort. Leather is soft, flexible, and adapts to the shape of the wrist over time, making it especially pleasant to wear for long periods. Compared to metal bracelets, leather straps are lighter and do not feel cold or hot against the skin, which improves wearing comfort in varying temperatures. Esthetic versatility is another major advantage: leather straps are available in a wide range of colors, textures, and finishes, allowing them to complement both elegant dress watches and casual or vintage styles. In addition, high-quality leather develops a unique patina over time, which many wearers find attractive and characterful.

On the downside, leather straps are generally less durable than metal or rubber alternatives. They are sensitive to moisture, sweat, and prolonged exposure to water, which can cause the leather to deteriorate, discolour, or develop odours. As a result, leather straps are not ideal for sports, swimming, or very hot and humid conditions. They also require more care, such as regular cleaning and occasional conditioning, to maintain their appearance and longevity. Unlike metal or synthetic straps, leather

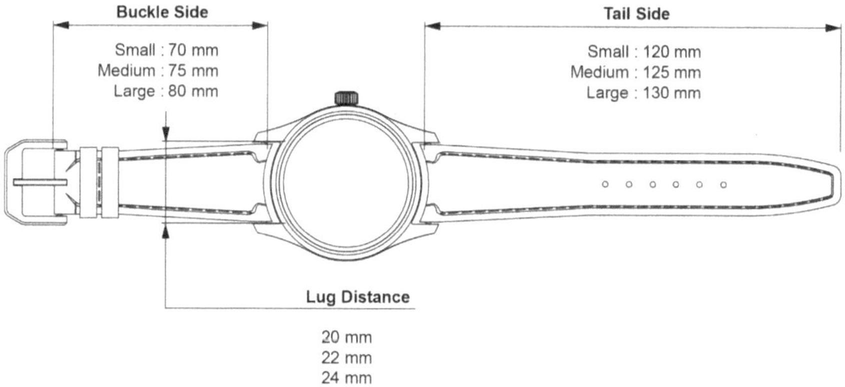

Fig. 2.41 Sizes of a wristwatch strap

bands wear out over time and usually need to be replaced periodically. Furthermore, leather straps may cause discomfort or skin irritation for some people with allergies or sensitive skin. While often perceived as elegant and traditional, leather straps can be less practical for everyday or highly active use compared to more robust materials.

There are quality grades for leather watch straps. These differences can affect durability, comfort, esthetics, and price. Here are the main factors that determine the quality of leather watch bands [66].

(a) **Type of Leather**

Full-Grain Leather

- highest quality
- made from the top layer of the hide
- retains the natural grain
- most durable
- develops patina

Top-Grain Leather

- slightly lower quality
- sanded or buffed to remove imperfections
- less durable than full grain

Genuine Leather

- marketing term for lower-quality leather
- made from leftover layers

Bonded Leather

- lowest grade
- made from leather scraps bonded together with adhesives
- cheap and not long-lasting

(b) **Tanning Method**

Vegetable Tanning

- traditional and natural method using plant tannins
- produces a rich patina over time
- has a distinct leather smell
- usually more expensive

Chrome Tanning

- uses chemicals (mostly chromium salts)
- faster and cheaper, but less environmentally friendly
- leather is softer but doesn't age as gracefully

(c) **Construction Quality**

Stitching

- High-quality straps have tight, consistent stitching.
- Hand-stitched straps are usually more premium.

Edge Finishing

- High-end straps have carefully painted or folded edges.

Lining

- Quality straps often have water-resistant linings for durability.

The following leather types are commonly used for watch straps:

(a) **Animal-Based Leathers**

- Calfskin—the most common; smooth, durable, and versatile
- Cowhide—thicker and more robust than calfskin
- Goatskin—light, flexible, with a fine grain
- Sheepskin/Lambskin—very soft and supple, less durable
- Pigskin—breathable with a distinctive grain pattern
- Deerskin—extremely soft and flexible

(b) **Exotic Leathers**

- Alligator leather—luxury leather with large, irregular scales
- Crocodile leather—similar to alligator, finer and more symmetrical scales
- Lizard leather—small, tight scales with a glossy finish
- Ostrich leather—recognizable by quill follicles, very durable
- Stingray (Shagreen)—extremely tough, pebbled texture
- Snake leather (Python)—soft with large, distinctive scales

(c) **Specialty and Processed Leathers**

- Shell Cordovan—dense, glossy horsehide leather, very durable
- Suede—napped leather with a soft, matte surface
- Nubuck—lightly sanded leather, velvety texture
- Pull-up leather—treated to show color variation when bent
- Vegetable-tanned leather—tanned with natural tannins, develops patina

(d) **Alternative and Modern Options**

- Vegan leather (synthetic or plant-based)—leather-like materials without animal origin
- Recycled leather—made from bonded leather fibres

Certain watch straps made from exotic materials—especially alligator, crocodile, lizard, ostrich, or snake leather—are subject to strict international wildlife protection laws. These regulations can limit or even prohibit the sale, import, export, or transport of such straps in specific countries.

The most important legal framework is the international agreement known as CITES (*Convention on International Trade in Endangered Species of Wild Fauna and Flora*). More than 180 countries participate in this treaty. CITES regulates trade in animal and plant species that could become endangered through uncontrolled commercial exploitation.

In the watch industry, this mainly affects straps made from:

- Alligator leather
- Crocodile leather
- Certain lizard species
- Python or snake leather

Different species are categorized under different CITES appendices:

- **Appendix I** species are highly endangered and generally cannot be traded commercially.
- **Appendix II** species may be traded legally, but only with official permits and traceable documentation. Most legally used alligator and crocodile leathers fall into this category.

For example, American alligator leather is generally legal because the species is considered sustainably managed and farm-raised. However, every hide must be documented and traceable through official CITES certification.

Many countries impose additional national restrictions beyond CITES:

- The European Union applies especially strict import regulations and often requires import permits even for species that are only moderately protected under CITES.
- The United States regulates imports through the U.S. Fish & Wildlife Service and customs authorities. Incorrect paperwork can lead to confiscation of goods.
- Some countries prohibit specific species entirely, regardless of documentation. For example, trade involving Siamese crocodile leather from certain Southeast Asian countries is heavily restricted or forbidden.

These legal rules can create several practical consequences for watch manufacturers and retailers:

- Watches with exotic leather straps may not be shipped internationally without permits.
- Customs authorities may seize products lacking proper CITES documentation.
- Online marketplaces sometimes ban listings containing protected animal materials.
- Some brands remove exotic straps before exporting watches to certain regions.

As a result, many watch companies now offer alternative strap materials such as:

- Calf leather with alligator embossing
- Rubber straps
- Textile straps
- Vegan leather alternatives

Luxury manufacturers such as IWC Schaffhausen explicitly state that their alligator leather complies with CITES requirements and originates from regulated farms [67].

2.3.2 *Textile Straps*

One of the textile straps' main benefits is comfort. Textile straps, such as nylon, canvas, or fabric NATO straps, are lightweight, flexible, and breathable, which makes them especially comfortable in warm weather and suitable for long periods of wear. They adapt easily to the wrist and usually dry faster than leather, making them a practical option for casual daily use and outdoor activities. Another advantage is style versatility: textile straps are available in many colors, patterns, and weaves, allowing wearers to easily change the look of a watch at a relatively low cost. They are also generally more affordable than metal bracelets or high-quality leather straps and are easy to replace.

However, textile straps also have disadvantages when compared to other materials. They tend to be less durable than metal bracelets and can show wear, fraying, or fading over time, especially with frequent washing or heavy use. Textile straps can absorb sweat, dirt, and odours, which may require regular cleaning to maintain hygiene and appearance. While they are more water-tolerant than leather, they can remain damp against the skin for some time after getting wet, which may be uncomfortable. From an esthetic standpoint, textile straps are usually perceived as more casual and less refined, making them less suitable for formal or business settings. In addition, they often provide less structural support and a less premium feel than metal or leather alternatives.

Textile watch straps are often paired with leather, rubber, or synthetic lining on the underside for comfort and sweat resistance. They are typically made from these materials [68, 69]:

(a) **Nylon**

- Synthetic fibre

Synthetic fibre.

Advantages

- Very durable and lightweight.
- Water- and sweat-resistant.
- Quick-drying.
- Affordable and widely available.
- Comfortable once broken in

Disadvantages

- Can feel stiff at first.

- May fray over time if not heat-sealed at the edges.
- Less formal in appearance.

(b) **Canvas**

Strong woven cotton or cotton-blend fabric.

Advantages

- Rugged, casual look.
- Breathable and comfortable against the skin.
- Affordable and versatile.

Disadvantages

- Absorbs sweat and water.
- Can stain easily.
- Wears out faster than synthetics.

(c) **Polyester**

Synthetic fibre.

Advantages

- Strong and abrasion-resistant.
- Retains colors well (less fading than nylon or cotton).
- Lightweight and quick-drying.
- Smooth texture, comfortable on skin.

Disadvantages

- Less natural feel compared to cotton canvas.
- Can trap heat and feel less breathable.

(d) **Sailcloth**

Tightly woven synthetic fabric (often polyester or nylon).

Advantages

- Extremely tough and durable.
- Resistant to UV, water, and abrasion.
- Stylish, often used in luxury sports watches.
- Retains shape well.

Disadvantages

- Stiffer than nylon or canvas at first.
- Usually more expensive.
- May require lining for comfort.

(e) **Denim**

Cotton twill fabric, sometimes used for fashion-oriented straps.

Advantages

- Unique, casual, fashion-oriented style.
- Comfortable once broken in.
- Ages with character (like jeans).

Disadvantages

- Not water-resistant.
- Prone to fading and staining.
- Less durable than nylon or sailcloth.

(f) **Kevlar®**

High-strength aramid fibre.

Advantages

- Very high strength-to-weight ratio.
- Resistant to cuts, abrasion, and heat.
- Sporty and technical look.
- Often paired with padding/lining for comfort.

Disadvantages

- Expensive compared to standard fabrics.
- Can feel stiff without lining.
- Not inherently water-resistant (usually treated).

2.3.3 Plastic and Rubber Straps

One of the main strengths of plastic straps and rubber straps is their high resistance to water, sweat, and humidity. Unlike leather or textile straps, plastic and rubber do not absorb moisture, making them ideal for swimming, sports, and outdoor activities. They are generally very durable, flexible, and resistant to environmental influences such as UV light and temperature changes. In addition, these straps are lightweight and easy to clean, which contributes to a high level of wearing comfort during active use. Their elasticity also allows for a secure fit on the wrist, even during movement. From a design perspective, plastic and rubber straps are available in many colors and shapes and often complement modern, sporty, or technical watch designs.

On the negative side, plastic and rubber straps are often perceived as less elegant or less premium than metal or leather alternatives, which limits their suitability for formal or business environments. Depending on the quality of the material, they may attract dust and lint or develop an unpleasant smell over time if not cleaned regularly. Some lower-quality rubber or plastic straps can harden, crack, or lose flexibility as they age. In hot conditions, these materials may also trap heat and reduce breathability, leading to sweat buildup and discomfort. Furthermore, compared to leather or metal bracelets, plastic and rubber straps usually offer less refinement in terms of texture and aging character, as they do not develop a patina and tend to look worn rather than characterful over time.

The materials differ slightly in their properties and applications.

(a) **Plastic (synthetic) straps**

Materials used

- Polyurethane (PU).
- Polyvinyl chloride (PVC).
- Resin-based plastics.

Advantages

- Lightweight and comfortable to wear.
- Resistant to water and sweat.
- Very affordable.
- Available in many colors and designs.

Disadvantages

- Can feel less premium.
- May become brittle and crack over time, especially under UV-light or extreme temperatures.
- Less breathable, which can cause sweating under the strap.

(b) **Rubber straps**

Materials used

- Natural rubber (less common today).
- Synthetic rubbers such as silicone or fluorocarbon rubber.

Advantages

- Very flexible and soft, providing excellent comfort.
- High water resistance, making them ideal for diving and sports watches.
- Durable against chemicals, oils, and saltwater (especially high-quality FKM rubbers).

Disadvantages

- Natural rubber can degrade, becoming sticky or brittle over time.
- Silicone straps attract dust and lint easily.
- High-quality synthetic rubbers (like FKM) are more expensive than plastic or silicone.

2.3.4 *Sustainable Straps*

There is a growing demand for sustainable watch straps because consumers are becoming more aware of the environmental and social impact of their purchasing decisions. Many people want to reduce waste, avoid animal-based or petroleum-derived materials, and support brands that act responsibly and transparently. At the same time, sustainable materials have improved significantly in quality, durability, and comfort, making them a practical alternative to traditional leather and plastic straps.

(a) **Recycled Materials**

These use post-consumer waste, reducing landfill and raw material extraction.

Recycled PET

- Made from recycled plastic bottles.
- Durable and water-resistant.

Recycled Nylon (e.g., ECONYL®)

- Regenerated from fishing nets and other nylon waste.
- Strong, flexible, and often used in NATO or fabric straps.

(b) **Biobased and Natural Fibers**

Derived from plants or natural sources—often compostable or low-impact.

Organic Cotton

- Grown without harmful pesticides; breathable and soft.
- Often woven into fabric straps.

Hemp

- Extremely durable and low impact to cultivate.
- With use, it softens and gains character.

Linen

- Made from flax plant fibres.
- Lightweight, breathable, and biodegradable.

(c) **Sustainable Plant-Based Leathers**

Materials designed to resemble leather without animal products.

Cork Leather

- Harvested from cork oak bark without harming the tree.
- Light, water-resistant, and unique in texture.

Pineapple Leather (Piñatex™)

- Made from pineapple leaf fibres—a by-product of agriculture.

Apple Leather

- Produced from apple peels and pomace left over in juice production.
- Offers a smooth, leather-like finish.

Mushroom Leather (e.g., Mylo™)

- Derived from mycelium (fungal root structures).
- Biodegradable and highly customizable.

Cactus Leather

- Made from mature cactus leaves.
- Durable and supple with a low water footprint.

(d) **Vegetable-Tanned and Eco-Treated Leathers**

Not vegan, but more sustainable than traditional leather.

Vegetable-Tanned Leather

- Uses plant tannins instead of chrome and heavy chemicals.

Chrome-Free/Low-Impact Leather

- Tanned with fewer toxins and improved wastewater management.

2.3.5 NATO Straps

NATO straps are a specific type of textile watch strap that originated in a military context and are valued for their practicality, durability, and distinctive appearance. The name "NATO strap" comes from their adoption by the British Ministry of Defence in the 1970s, where they were issued to soldiers under a NATO stock number. These straps were designed to be simple, robust, and reliable under demanding conditions.

A NATO strap (see Fig. 2.42) is typically made from woven nylon or similar synthetic fabric and is constructed as a single, continuous piece that passes underneath the watch case. A secondary strip of material runs beneath the case back, creating a double-layer design. One of the key advantages of this construction is security: if one spring bar fails, the watch head remains attached to the wrist by the other bar, significantly reducing the risk of losing the watch. The strap is usually secured with a pin buckle and multiple metal keepers, allowing for a wide range of adjustment.

In terms of comfort, NATO straps are lightweight, flexible, and breathable, making them well-suited for everyday wear, warm climates, and active use. They are also highly resistant to water and dry relatively quickly compared to leather straps. From a stylistic standpoint, NATO straps are available in a vast range of colors, patterns,

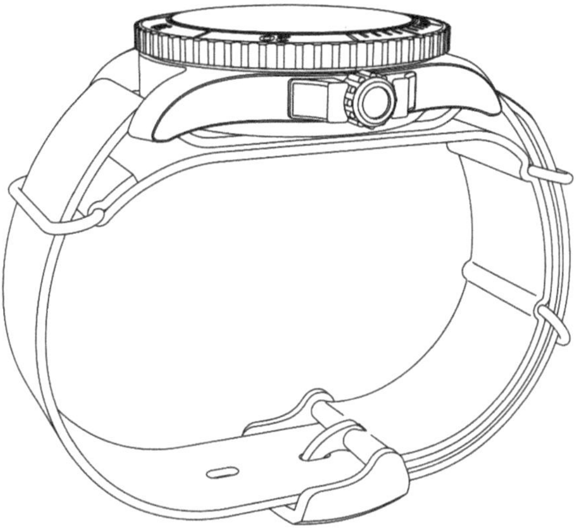

Fig. 2.42 Watch with a NATO-strap

and stripe combinations, from classic military designs to bold and modern variations, allowing wearers to easily change the character of a watch.

However, NATO straps also have some drawbacks. Because the strap passes underneath the watch case, it raises the watch slightly on the wrist, which may be uncomfortable or aesthetically unappealing for some users, especially on thicker watches. Additionally, NATO straps are generally perceived as casual and utilitarian, making them less suitable for formal or business settings. Despite these limitations, NATO straps remain popular for their versatility, affordability, and strong association with military and tool-watch heritage [70].

2.3.6 Bracelets

Metal watch bracelets offer several advantages over other types of watch straps, but they also come with some notable disadvantages. One of the main benefits of metal bracelets is their durability. Made from materials such as stainless steel or titanium, they are highly resistant to wear, tearing, and environmental influences, and they generally last much longer than leather, textile, or rubber straps. Metal bracelets are also well-suited for daily use and for exposure to water, sweat, and temperature changes, making them a popular choice for sports and tool watches. In addition, they require relatively little maintenance, as they can be cleaned easily and do not absorb moisture or odours. From an esthetic perspective, metal bracelets often give a watch a solid, high-quality appearance and can enhance the perceived value of the timepiece. However, metal bracelets also have disadvantages when compared to other strap types. They are usually heavier, which can reduce wearing comfort, especially for people with smaller wrists or those who prefer lightweight watches. Metal can also feel cold on the skin in low temperatures and hot in direct sunlight, which may be uncomfortable. Adjusting the size of a metal bracelet often requires tools or professional assistance, whereas leather or textile straps can be adjusted quickly using a buckle. Furthermore, metal bracelets are more prone to visible scratches over time, particularly on polished surfaces. Finally, they tend to be more expensive than many alternative straps and offer less flexibility in terms of color, texture, and casual styling.

Figure 2.43 gives an overview of the different types of metal watch bracelets, grouped by construction, style, and material [71].

(a) **Types of Metal Watch Bracelets (by Design and Construction)**

Oyster Bracelet

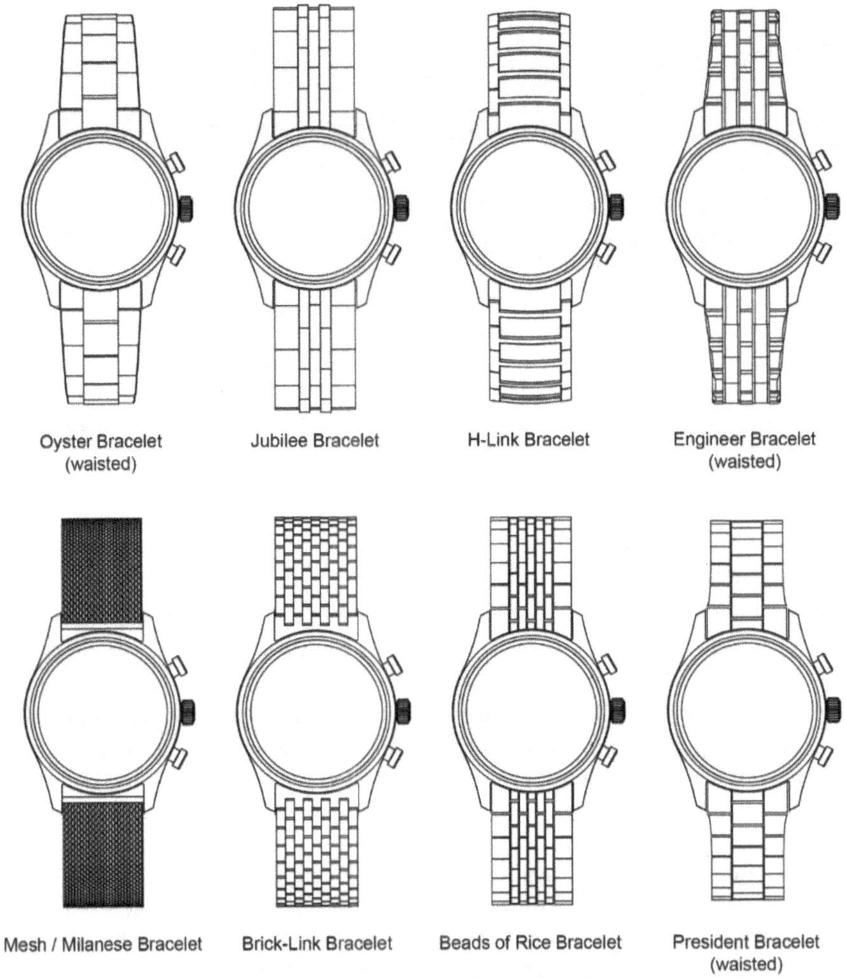

Fig. 2.43 Popular types of bracelets

- Flat, wide three-link design
- Very robust and comfortable
- Commonly used on sports and tool watches

Jubilee Bracelet

- Five-link design with smaller, rounded center links
- More flexible and elegant than the Oyster
- Often found on dress-sport watches

H-Link Bracelet

- Links shaped like the letter "H"
- Clean, modern, and structured look
- Often used on contemporary sports watches

Engineer Bracelet

- Thick, solid links with a technical appearance
- Heavy and highly durable
- Common on professional or tool watches

Mesh/Milanese Bracelet

- Made of finely woven metal wire
- Very breathable and adjustable
- Originally developed in Italy

Brick-Link Bracelet

- Rectangular, closely packed links
- Vintage-inspired design
- Creates a bold, architectural appearance

Beads of Rice Bracelet

- Multiple small, rounded links resembling grains of rice
- Extremely flexible and comfortable
- Popular on vintage and dress watches

President Bracelet

- Semi-circular three-link design
- Combines comfort and luxury
- Typically associated with high-end watches

Watch bracelets can have a straight profile; in this case, they have the same width between the lugs as at the clasp. If, on the other hand, they taper towards the clasp, they are called waisted bracelets.

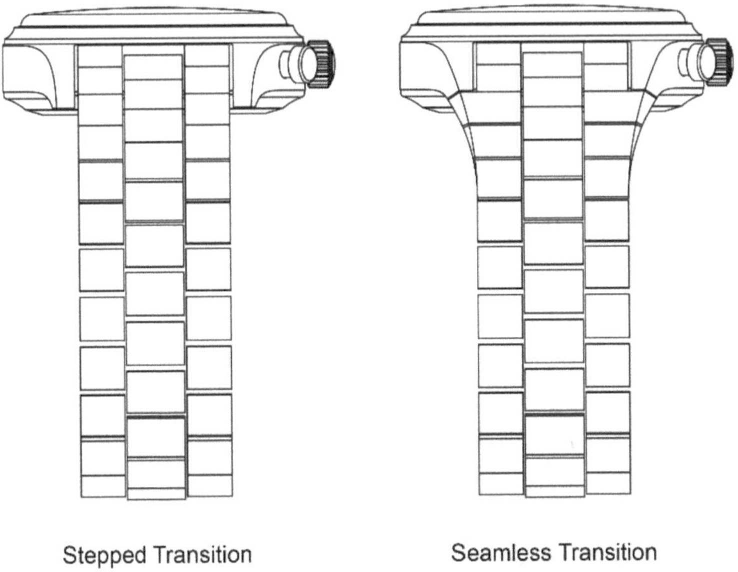

Fig. 2.44 Stepped and seamless watch bracelet transition

The ends of most watch bracelets are between the case lugs. This creates a stepped transition to the watch case. Some watches, however, feature bracelets whose sides blend seamlessly into the sides of the case. The difference is presented in Fig. 2.44.

(b) **Types of Metal Watch Bracelets (by Material)**

Stainless Steel

- Most common material
- Corrosion-resistant and durable

Titanium

- Lightweight and hypoallergenic
- Excellent strength-to-weight ratio

Gold (Solid or Plated)

- Used in luxury watches
- Warm tone and prestigious appearance

Two-Tone (Steel and Gold)

- Combines stainless steel with gold elements
- Balances sportiness and elegance

Platinum

- Extremely rare and heavy

- Reserved for ultra-luxury timepieces
- **Finishing**
- Brushed finish—matte, tool-watch appearance
- Polished finish—glossy, dress-oriented look
- Combination finish—brushed and polished surfaces for contrast

2.3.7 Closing-Systems

A watch clasp is an essential functional element of every wristwatch. Its construction is crucial for how comfortably a watch can be put on and how securely it fastens to the wrist. Its design can either enhance or detract from the elegance of a watch. The most important types of locking mechanisms include pin buckles, deployant clasps, and butterfly clasps. These buckles all have in common that they can be operated with one hand [72, 73].

(a) **Pin Buckle**

A pin buckle is the most traditional and simplest type of watch strap closure, similar to a belt buckle. They are particularly suitable for casual fabric straps and NATO designs.

How it works:

- A metal pin goes through one of the holes in the strap.
- The excess strap is tucked into one or two strap keepers.

Advantages:

- Simple, lightweight, and reliable.
- Easy to adjust precisely.
- Often found on leather, rubber, or fabric straps.

Disadvantages:

- Causes more wear on the strap holes over time.
- Requires fully opening the strap when putting the watch on or taking it off.

(b) **Deployant Clasp**

A deployant clasp is a folding clasp designed to reduce strap wear and improve comfort and security. These clasps are found especially on premium leather straps. Folding clasps are also offered with a mechanism that allows for fine adjustment of the strap width. This fine adjustment is advantageous if the wrist circumference increases, for example, on hot days.

How it works:

- The strap opens via two folding metal blades.
- The watch is put on without fully separating the strap.

- Often combined with a push-button release or safety lock.

Advantages:

- More secure than a pin buckle.
- Reduces stress and wear on the strap.
- Easier and safer to put on and remove the watch.

Disadvantages:

- More complex and heavier than a pin buckle.
- Slightly less precise fit adjustment, depending on the design.
- Usually more expensive.

(c) **Butterfly Clasp**

A butterfly clasp is a specific type of deployment clasp characterized by its symmetrical, concealed design. Due to their more compact design compared to folding clasps, butterfly clasps are primarily found on women's watch straps. Their symmetrical design makes them an ideal complement to metal watch straps.

How it works:

- Two folding blades open outward like butterfly wings.
- When closed, the clasp is hidden, creating a clean and seamless look.
- Commonly used on metal bracelets and high-end leather straps.

Advantages:

- Elegant, symmetrical appearance
- Even weight distribution on the wrist
- Secure and comfortable once properly adjusted

Disadvantages:

- Can be cumbersome to use
- Can be harder to size correctly
- Often more expensive than standard deployment clasps

Figure 2.45 shows examples of closing systems for each group.

2.3.8 Attachment-Systems

In wristwatches, the strap or bracelet is usually connected to the watch case through the lugs (the protrusions on both sides of the case). Over time, several attachment systems have been developed, differing in security, ease of replacement, and design [74].

(a) **Spring Bars**

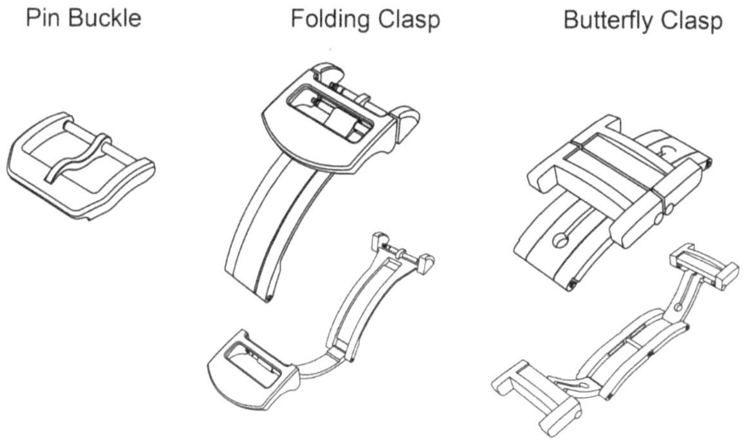

Fig. 2.45 Most common types of closing systems

The most widespread attachment method is the spring bar system. A spring bar is a small metal tube with spring-loaded pins at both ends. When compressed, the bar becomes slightly shorter and can be inserted between the lugs. Once released, the pins extend into holes in the lugs and lock the strap or bracelet in place.

(b) **Quick Release Spring Bars**

A modern evolution of the spring bar is the quick-release spring bar. These bars include a small lever or tab that can be operated with a fingernail. This allows the strap to be removed and installed without a dedicated spring-bar tool. They reduce the risk of scratching the wristwatch when changing the strap.

(c) **Screw Bars**

Some watches use screw bars instead of spring-loaded bars. This very secure system is favoured for some dive watches, military-style watches, and luxury brands.

(d) **Integrated Bracelets and Proprietary Systems**

Many modern watches use integrated bracelets or manufacturer-specific attachment systems. In these designs, the bracelet or strap blends directly into the case shape. Standard spring bars may be absent, and replacement straps often need to be made specifically for that watch model. Some manufacturers have developed proprietary quick-change mechanisms that allow tool-free strap swaps while maintaining a seamless appearance.

2.4 Movement

In high-end mechanical watchmaking, the esthetic design of a movement goes far beyond pure functionality. While all components must work precisely together, how they *look*—particularly when visible through a transparent case back or in skeletonized models—is a crucial part of the movement's appeal.

The movement often defines the proportions of the case, the thickness of the watch, and even the layout of the dial.

Figure 2.46 shows the main elements of a hand-wound mechanical movement.

2.4.1 Bridge Shapes

In a mechanical movement, bridges and plates are structural components that hold the gears, wheels, escapement, and other moving parts in place. They are fixed to the main plate and ensure correct alignment and stability. As watch movement design evolved, watchmakers developed different ways to organize these supports—from large plates covering much of the movement to individual bridges holding single components. Figure 2.47 shows the differences.

Full Bridge

A full bridge is a large structural piece that spans a significant portion of the movement's top side, often covering the entire width of the movement. It supports multiple components—such as the gear train and barrel—under one solid piece rather than several smaller bridges. This design offers high stability and rigidity, helping maintain precise alignment over time. A full bridge also provides a large surface for esthetic finishing (engraving, polishing), which is appreciated in high-end movements [75].

Three-Quarter Bridge

The three-quarter bridge covers about three-quarters of the movement's width, typically leaving only a portion (often where the balance wheel is located) uncovered or supported by a separate small cock. It therefore still supports most of the gear train and barrel, but allows more visibility of individual components than a full bridge. This style is a balance between structure and visibility. It is especially known from German watchmaking (e.g., A. Lange & Söhne), where it became a hallmark due to its robust stability and ease of assembly compared to older craft layouts with many tiny bridges [76].

Half Bridge

A half bridge is smaller, typically covering only part of a wheel or train—roughly half of the width that a full piece would occupy. It supports fewer components and is often used in designs that emphasize visibility and openness of the movement. Since half bridges leave more of the underlying architecture visible, they are often

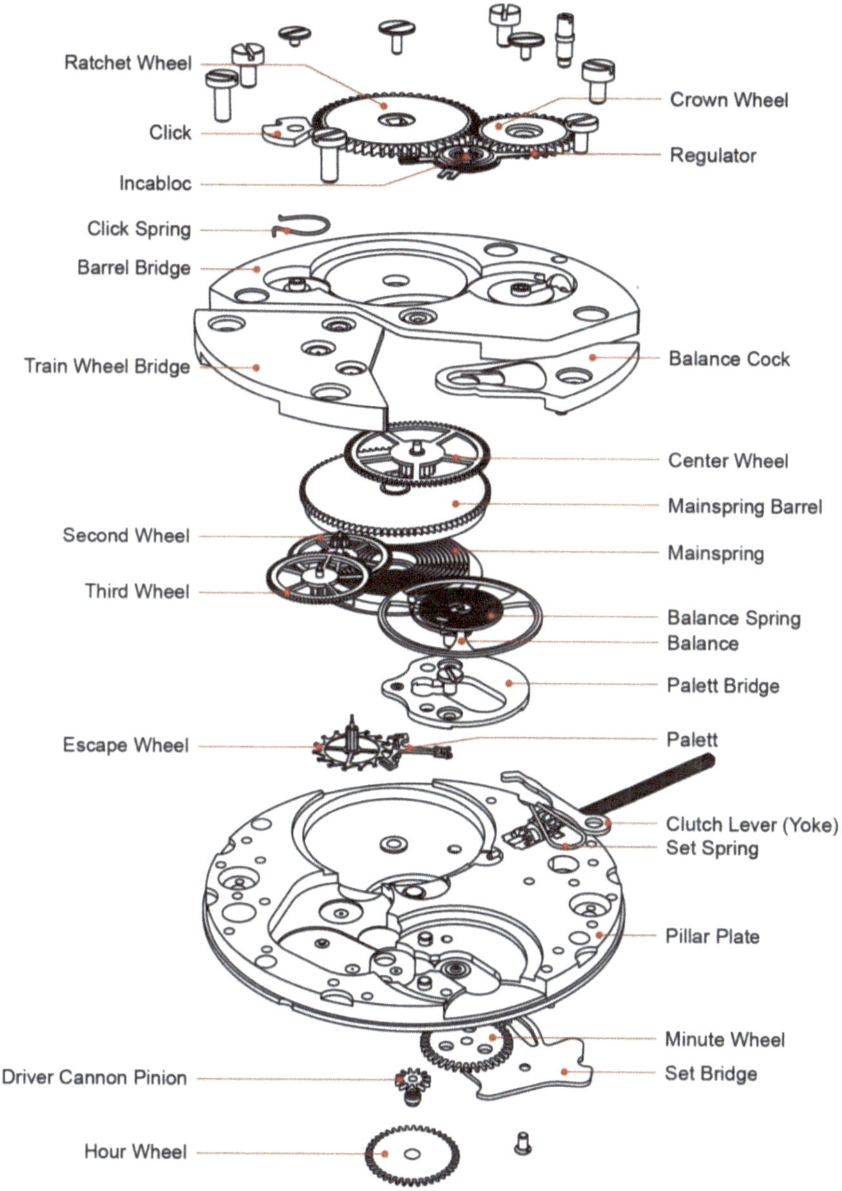

Fig. 2.46 Elements of a mechanical watch movement shown in an exploded view

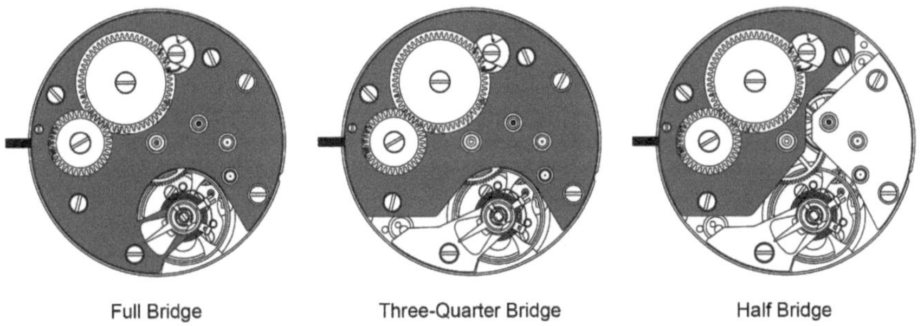

Fig. 2.47 Different bridge shapes

found in skeletonized or open-worked watches where the visual presentation of the mechanism is a priority. Like full and three-quarter bridges, half bridges can be finished or decorated to contribute aesthetically.

2.4.2 Bridge Surfaces

The primary esthetic consideration is finishing (often called *finissage*), which encompasses a wide array of surface treatments (as shown in Figs. 2.48, 2.49, 2.50, 2.51, 2.52, 2.53, 2.54 and 2.55) applied to the movement's components. Finishing not only enhances visual appeal but also reflects the watchmaker's skill and the piece's level of refinement [77–79].

Black Polish Finishes

- Surfaces polished to reflect light in specific ways
- Creating visual contrast and a sense of luxury

Fig. 2.48 Movement bridges with a black polish finishing

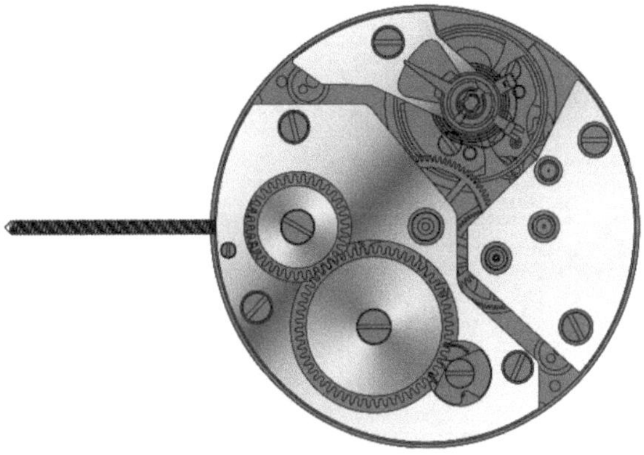

Fig. 2.49 Movement bridges with a linear brush finishing

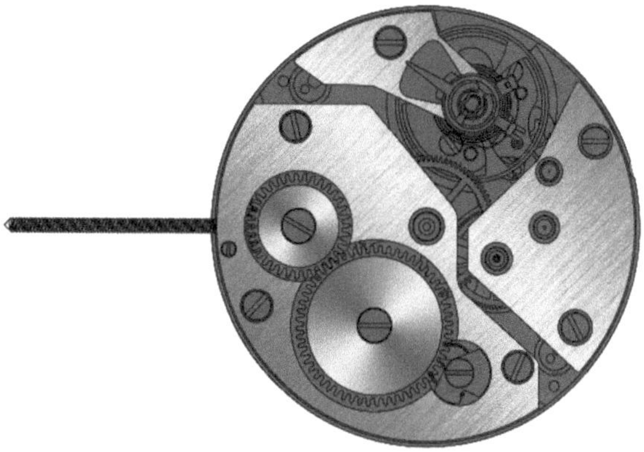

Fig. 2.50 Movement bridges with a perlage finishing

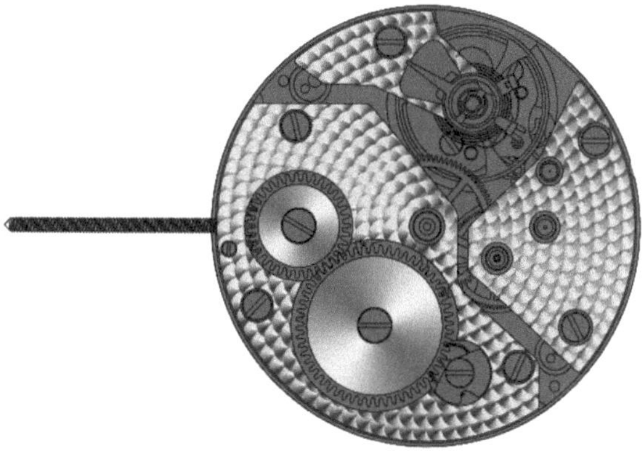

Fig. 2.51 Movement bridges with a linear Côtes de Genève finishing

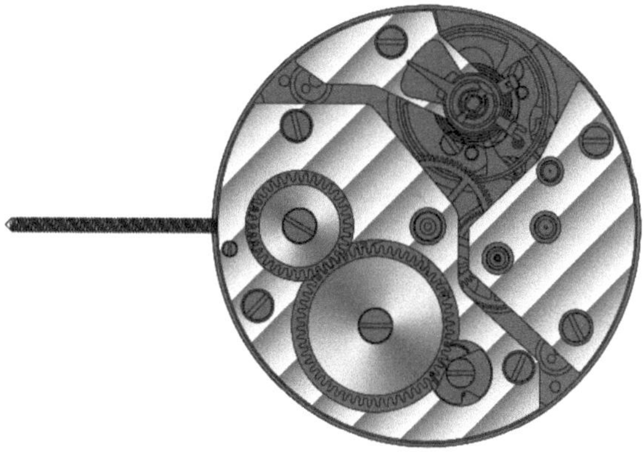

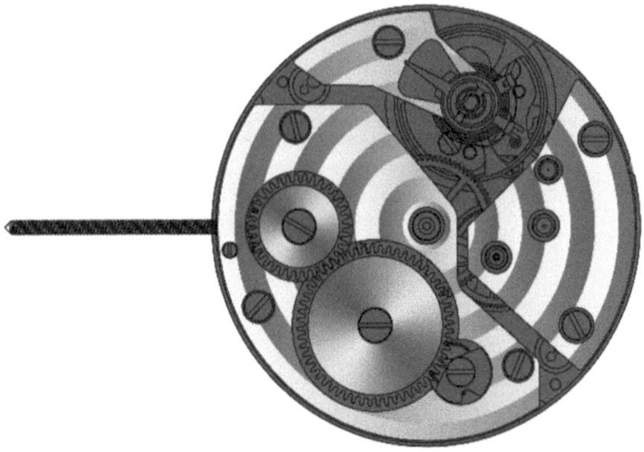

Fig. 2.52 Movement bridges with a circular Côtes de Genève finishing

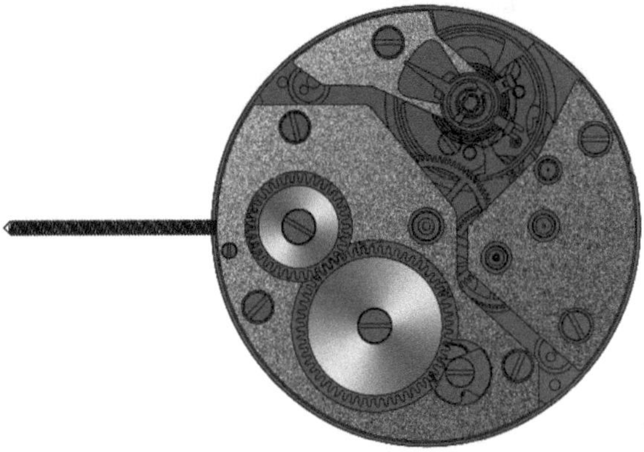

Fig. 2.53 Movement bridges with a frosting finishing

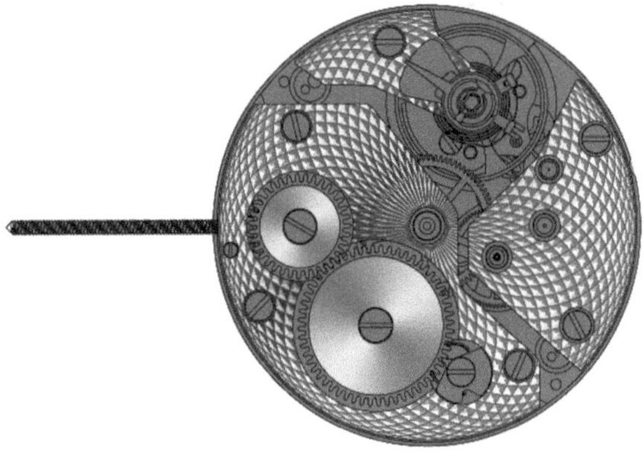

Fig. 2.54 Movement bridges with a Guilloché finishing

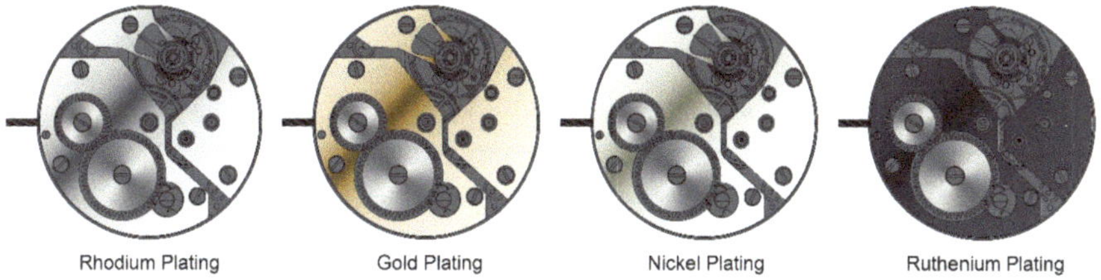

Fig. 2.55 Movement bridges with an electrochemical colouring

Brush Finishes

- Linearly or circularly brushed surfaces
- Creating a technically modern look

Perlage (Circular Graining)

- Overlapping circular patterns that add texture and depth.
- Often on main plates or recessed areas.

Geneva Stripes (Côtes de Genève)

- Elegant, symmetrical appearance.
- Create light play across the movement.
- Signature of classic haute horlogerie.

Frosting

- A matte finish is applied to the surface of components, often using a chemical or mechanical process.
- Used to create a contrast to shiny surfaces.

Guilloché

- Engraving technique.
- Patterns require a special engine turning machine.

Electroplating is an electrochemical process in which a thin metallic coating is deposited onto a brass component by passing an electric current through an electrolyte solution containing metal ions. The brass part acts as the cathode (negative electrode), causing metal ions from the plating bath to be reduced and deposited on its surface.

The final color is determined by the type of metal or alloy deposited. Besides providing decorative coloration (see Fig. 2.55), the electroplated layer can improve corrosion resistance, wear resistance, and surface hardness.

In watchmaking, brass bridges and plates are often polished and decorated before electroplating so that the deposited coating preserves and enhances the underlying surface finish.

Rhodium Plating

- Bright silver-white color

- Excellent corrosion resistance
- High hardness and wear resistance
- Frequently used in contemporary luxury movements

Gold Plating

- Prestigious appearance
- Good corrosion resistance
- Can be adjusted to different color tones through alloy chemistry

Nickel Plating

- Bright silver appearance
- Lower cost than rhodium
- Good corrosion resistance
- Can be polished to a high luster

Ruthenium Plating

- Dark grey to black appearance
- Excellent wear resistance
- Modern, technical visual style
- Often combined with skeletonized movements

Anglage (Chamfering/Bevelling)

- Edges of parts are bevelled and polished to a mirror finish (see Fig. 2.56), emphasizing shape and depth.
- This painstaking handwork is highly valued by collectors.

Esthetic movement design often plays with *contrast and texture*—for example, combining matte surfaces with mirror polish, or juxtaposing perlage with linear Geneva stripes. These contrasts enhance the perception of depth and complexity when viewed through a crystal case back. The thoughtful use of different textures

Fig. 2.56 Movement with chamfered bridges

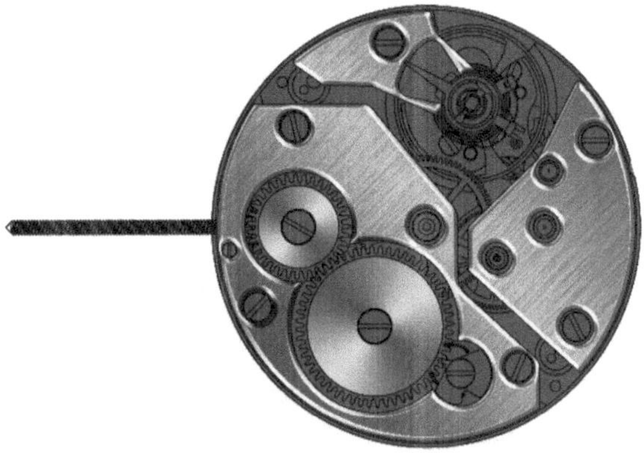

makes the movement more visually engaging and allows light to interact with each surface in different ways.

Movement esthetics also help express a brand's identity or design philosophy. For example, some manufacturers emphasize minimal, Bauhaus-like movement layouts, while others celebrate traditional decorative arts. This stylistic language can differentiate a timepiece and give it a unique character. Ultimately, the esthetic design of a mechanical movement is about emotional expression blended with technical precision. It invites the viewer to appreciate not just how a watch *functions*, but how it *feels*—celebrating the craftsmanship, the meticulous handwork, and the artistry that elevate mechanics into horological art.

2.4.3 Engravings

Several types of information are commonly engraved or stamped on bridges (or on nearby plates) during manufacture [80]:

- Manufacturer/Name or Logo
- Caliber or Movement Number
- Number of jewels
- Country of Manufacture or Origin
- Certification or Quality Marks
- Decorative Elements

Figure 2.57 illustrates engravings on a hand-wound movement.

The most well-known quality seals found on watch movements include.

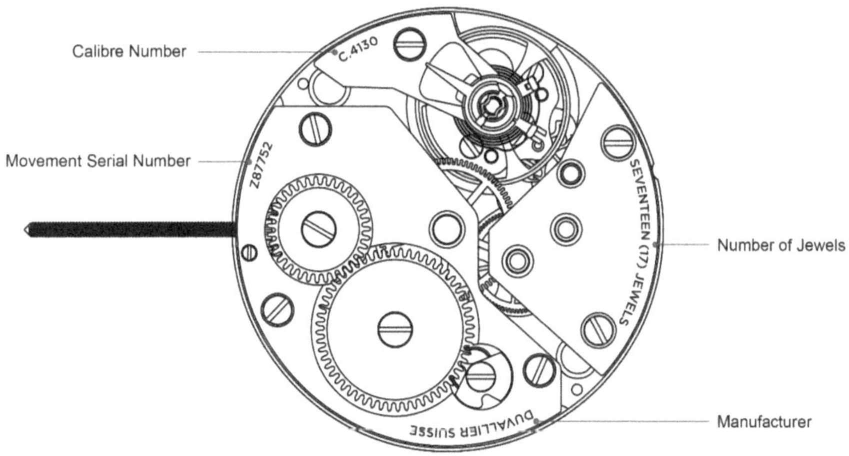

Fig. 2.57 Watch movement with usual engravings

The COSC certificate

Refers to a chronometer certification issued by the Contrôle Officiel Suisse des Chronomètres (COSC), the official Swiss chronometer testing institute. It is an independent, non-profit organization responsible for testing and certifying the precision of Swiss-made watch movements.

A watch movement that receives COSC certification is officially designated as a chronometer, meaning it has met strict accuracy and performance standards under controlled laboratory conditions [81].

The Geneva Seal

—or *Poinçon de Genève*—is one of the most prestigious quality hallmarks in traditional Swiss watchmaking. It certifies that a mechanical movement has been manufactured to the highest standards of craftsmanship, origin, finishing, and performance according to strict regulations overseen by the Office for the Voluntary Inspection of Watches in the Canton of Geneva. Only mechanical watch movements are eligible for the Geneva Seal. A fundamental requirement is that the movement's assembly and adjustment must be carried out inside the City or Canton of Geneva [82].

The movement must meet a detailed set of technical and esthetic criteria defined in the Seal's regulations. These criteria relate to both quality of workmanship and finishing:

- Hand-finished components with no visible machining marks.
- Bevelled and polished edges on parts and screws.
- Polished jewel holes with chamfered settings.
- Geneva stripes (*Côtes de Genève*) and circular graining where required.
- Polished steel parts and meticulously finished surfaces.

2.4.4 Rotor

In an automatic mechanical movement, the oscillating weight (rotor) is the component that winds the mainspring using the motion of the wearer's wrist. Figure 2.58 shows three types of rotors.

Full rotor

A semi-circular mass that rotates 360° around the center of the movement. It is efficient and robust, but partially covers the movement when viewed through a sapphire case back.

Micro-rotor

This version is smaller and integrated into the plane of the movement rather than mounted on top. This allows for a thinner movement and an unobstructed view of the finishing, but it is technically more complex and can be less efficient if not well engineered.

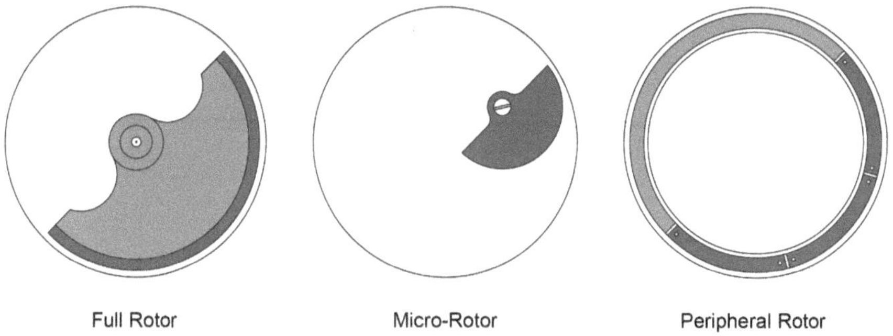

Fig. 2.58 Common rotor designs

Peripheral rotor

The rotor runs around the outer edge of the movement. This solution combines full winding efficiency with excellent visibility of the movement, but it requires high manufacturing precision and sophisticated bearing systems.

The rotor is often treated as a decorative focal point. Common finishing techniques include Geneva stripes, circular graining, sunburst brushing, perlage, engraving, and polishing. Brand logos, coats of arms, or inscriptions are frequently incorporated. In high-end watchmaking, rotors may be hand-engraved, black-polished, or even crafted from precious metals to emphasize exclusivity.

Many manufacturers use the rotor as a signature element. Distinctive shapes, cutouts, or finishes can immediately identify a brand or collection. In some cases, the rotor design echoes historical references, technical themes, or the overall esthetic of the watch [83].

2.5 Individualization

The possibilities for customizing a wristwatch according to customer requirements are extensive and allow for a highly individual product that reflects the wearer's personality, style, and specific needs.

Engraving

Individual engravings on wristwatches are common, especially when the watch is given as a gift or meant to mark a special occasion. These engravings are not legally required, but rather personal additions chosen by the owner or giver. They can include:

- Names and initials—of the wearer, gift giver, or both.
- Dates—such as birthdays, anniversaries, graduations, or wedding days.
- Short messages or dedications.

- Quotes or mottos—personal life mottos, religious phrases, or inspirational sayings.
- Symbols or motifs—zodiac signs, family crests.
- Special achievements—engravings to commemorate career milestones, awards, etc.

Strap

Another major area of customization is the watch strap, where clients can select from various materials such as leather, stainless steel, rubber, textile, or exotic leathers, as well as choose colors, stitching styles, and clasp types.

Dial

The dial design offers further possibilities, including custom colors, textures, numerals, indices, hand shapes, and even personalized logos or symbols.

Gemstone settings

In the luxury segment, customers may also request gemstone or diamond settings on the bezel, dial, or case to enhance exclusivity.

Additionally, bespoke manufacturers often allow modifications to the case material and finish, such as stainless steel, titanium, gold, or rose-gold coatings, along with polished, brushed, or matte surfaces. Some watchmakers even offer customization of the movement, for example, through engraved rotors, skeletonized components, or visible case backs that showcase the mechanics. Altogether, these options transform a standard wristwatch into a unique and personal timepiece tailored precisely to the customer's wishes.

Personalization is gaining increasing importance in the watch industry. This is because personalized products are increasingly perceived as a luxury. They serve to reflect the owner's identity. Furthermore, unique pieces have the potential to become sought-after collector's items.

Providers specializing in wristwatch personalization allow their customers to configure timepieces based on standard models online. Customers can select from a defined range of options, including case coatings, dial colors, printing, and straps.

Some renowned watch manufacturers are increasingly offering their customers the opportunity to influence the design of their products. VIP customers of some companies can collaborate with their designers to create unique watches [84].

Exercise 2

Solve the multiple-choice quiz. Find the correct completion for each sentence. Only one is correct in each case. You'll find the solutions at the end of the book.

2.1. Stainless steel...

a	☐	... cannot cause allergies.
b	☐	... is an alloy of iron and carbon.
c	☐	... has a melting point of 660 degrees Celsius.

2.2. Fluting...

a ☐ … is a process of milling parallel grooves or patterns on the watch case
b ☐ … is a process in which an electric current creates an oxide layer on a metal.
c ☐ … is a process to increase scratch resistance.

2.3. Acrylic …

a ☐ … is highly impact resistant.
b ☐ … cannot be polished.
c ☐ … is extremely scratch-resistant.

2.4. When small claws are used to attach gemstones to a watch, …

a ☐ … this is referred to as Channel-Setting.
b ☐ … this is referred to as Prong Setting.
c ☐ … this is referred to as Bezel Setting.

2.5. Cabochon crowns…

a ☐ … have a rounded insert set into the tip.
b ☐ … are placed on the left side of the case instead of the usual right.
c ☐ … are shaped like a cut diamond with multiple facets.

2.6. Guilloché Dials...

a ☐ … feature fine radial brushing from the centre outward.
b ☐ … transition from a lighter centre to darker edges.
c ☐ … have engraved geometric patterns, done by hand or engine-turning.

2.7. The so-called *California dial* …

a ☐ … combines both Roman and Arabic numerals on the same face.
b ☐ … divides the watch face into four clearly structured sections.
c ☐ … is made by firing powdered glass onto a base at high temperatures.

2.8. When a silicone pad transfers ink from an engraved plate onto the dial surface.…

a ☐ … this is referred to as screen printing
b ☐ … this is referred to as pad printing.
c ☐ … this is referred to as sublimation printing.

2.9. A Milanese bracelet…

a ☐ … is made of finely woven metal wire.
b ☐ … sports a five-link design with smaller, rounded centre links.
c ☐ … consists of a semi-circular three-link design.

2.10. The COSC certificate…

a ☐ … regulates trade in animal and plant species.
b ☐ … refers to a chronometer certification.
c ☐ … defines the requirements for divers' watches.

Chapter 3
Watch Design Workflow

Abstract Chapter 3 presents a systematic workflow for wristwatch design, from defining goals and preparing the design brief through research, concept development, rendering, and prototyping. It establishes key design principles—proportion, balance, and legibility—and examines information, redundancy, and brand archetypes. The chapter emphasizes market and historical research, brand-specific design codes, technical movement constraints, and iterative sketching. It then covers 2D and 3D visualization, 3D printing, and the preparation of precise technical drawings. Finally, it addresses manufacturing documentation, standards, tolerances, intellectual-property protection, confidentiality, and supplier responsibilities, highlighting the importance of disciplined planning and collaboration throughout the process.

A suitable introduction to the chapter on the design process is provided by an explanation of design principles relevant to watches and the distinction between information and redundancy.

The design process described below is ideal—straight out of a textbook. It contains no unforeseen problems that could disrupt the process to a greater or lesser degree. However, a watch designer should be prepared for problems. To avoid them, it is essential for the designer to stick strictly to an established schedule and to accept only binding decisions from his client or superior. A thorough working method is equally important. Time lost due to poor planning means lost revenue. A watch that launches a month later than planned only generates profit a month later.

Virtually all watches use purchased components. When selecting suppliers, reliability, clear communication, and adherence to deadlines are crucial.

3.1 Wristwatch Design Principles

Before discussing the actual design process, the most important principles in the design of a wristwatch will be highlighted.

G. Plange, *A Tutorial for Watch Designers*,
https://doi.org/10.1007/978-3-032-33722-1_3

Proportion

Proportion refers to the relative sizing of the different elements of the watch. The diameter of the case, the width of the bezel, the length and thickness of the lugs, and the size of the crown all need to be in harmony with one another and with the wrist it is intended to sit on. A watch with poorly considered proportions may look awkward—for example, a very small dial within a thick case, or oversized hands on a small dial. Good proportion ensures elegance and visual coherence, making the watch feel "right" at first glance.

Balance

Balance is about the visual distribution of elements across the dial and the case. A balanced watch design avoids any sense of heaviness or visual distraction on one side. This includes the placement of sub-dials, date windows, branding, and markers. For instance, chronograph sub-dials arranged symmetrically often look more harmonious than those unevenly placed. Balance can be symmetrical (mirror-like) or asymmetrical (deliberate contrast), but in both cases, the design must feel stable and well-composed.

Legibility

Readability, or legibility, is the ease with which the wearer can tell the time. This depends on the contrast between hands and dial, the clarity of markers, and the overall simplicity of the layout. A well-designed watch makes time instantly visible without straining the eyes. For example, high contrast (white hands on a black dial), well-sized indices, and uncluttered layouts all improve readability. Poor readability, on the other hand, can render even the most beautifully designed watch impractical [85].

3.2 Information and Redundancy

A watch is perceived by users as a message consisting of information and redundancy. The part of a message that is new to the recipient is called information. In contrast, redundancy represents the already known part. Anyone designing a product must decide on the information content conveyed. Limiting oneself to only a small amount of information creates a concise, but potentially banal, design. A large amount of information leads to complexity and can possess higher esthetic value. A complex industrial product allows users to continually discover new relationships of order within its design structure. A design with little information has fewer opportunities to ensure an aesthetic perception process in the long term.

Figure 3.1 shows a watch with an extremely large number of redundant elements. The clock can be perceived as either visually interesting or cluttered.

A watch that closely resembles the archetype of a wristwatch meets the expectations of users [86]. The Swiss psychologist Carl Gustav Jung coined the term

Fig. 3.1 Watch with many redundant elements (Courtesy of Casio)

archetype in his theory of the collective unconscious. In this context, archetypes are universal, symbolic patterns or primal images deeply rooted in the collective unconscious of all people. They influence our thinking, actions, and feelings. In product design, archetypes are used to build trust, generate emotional resonance, and enable intuitive use [87].

3.3 The Watch Design Process

The design process in the watch industry is not significantly different from that in other industries. It essentially involves creating the most attractive product (the wristwatch) possible based on existing or new technical components (the movement).

The process of designing a wristwatch is difficult to describe in general terms. The number of process steps and their duration depend primarily on the complexity of the product being designed. In professional design practice, a predefined plan for the design process is rarely found. The risk of limiting the designer's creativity through attempts to structure it is too great [88].

A design process can be divided into the steps of

- Goal definition
- Briefing
- Market and history research phase
- Exploration of brands' design codes
- Gathering of Technical Information
- Concept Phase
- Design Phase
- Prototype Phase

These steps are not always clearly defined. Loops are not only conceivable but common in this process. A negative evaluation of a design or prototype can lead to a change in the brief, which requires the creative process to be restarted. When scheduling a new product project, time buffers should therefore always be provided for additional design process loops [88].

The timeline shown in Fig. 3.2 is based on the author's experience. When a new movement is developed for a watch design, the process can take several years. From the prototype to the first produced watch, several more months typically pass.

3.4 Goal Definition and Briefing

A new watch design project always arises when a specific goal is to be achieved. The goal is usually to balance the difference between the actual state and a desired target state. This difference can relate to the expectations of consumers, retailers, or a manufacturer. Watch consumers are primarily concerned with the sheer enjoyment of the product, increased comfort, signalling a certain status, and much more. Retailers approach manufacturers with the desire for new models to increase their own sales. Major retailers often request exclusive special models to differentiate themselves from their competitors. A company's goals are usually based on monetary considerations. The goal is to maximize profits through the development of a new product to ensure the company's continued existence and to enable growth. The initiative to develop a new design usually comes from a manufacturing company.

A design brief is the foundation for the successful implementation of a design project. It ensures that all parties involved have a shared understanding of what is to be achieved. The brief (also called a specifications document or requirements specification) is prepared by the designer's client, who can be internal or external. It should always be in writing.

A good design briefing should include these details [89]:

Fig. 3.2 Usual timeline of a watch design process

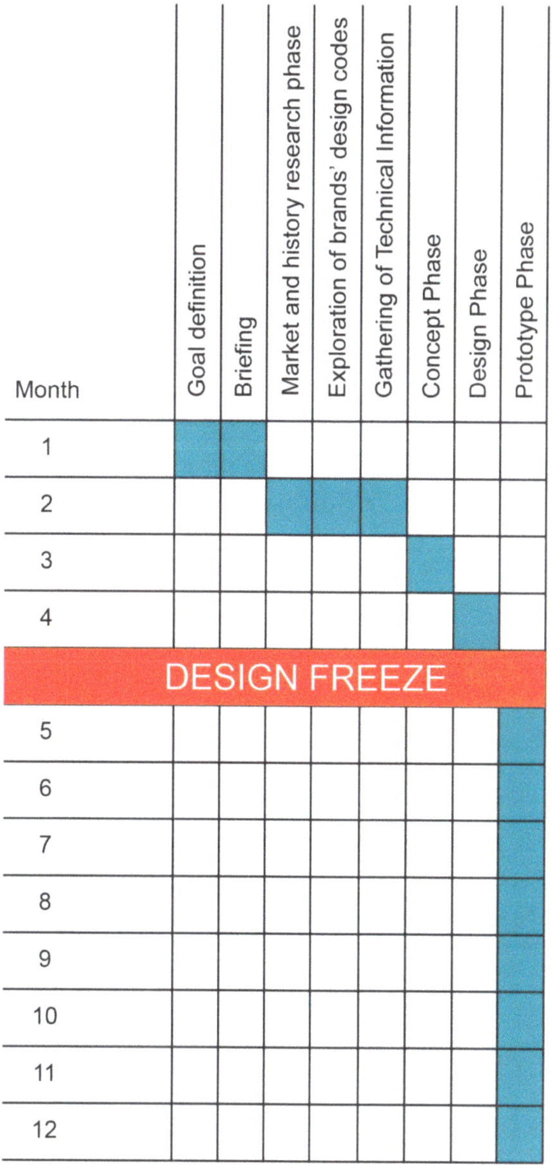

(a) **Technical Briefing**

The technical briefing concerns the product itself; it describes its classification in the manufacturer's product portfolio and its features. It should include the following points:

- Product family
- Watch movement (Manufacturer, Calibre number)
- Case material(s)
- Case diameter and distance of lugs
- Required water pressure resistance in Bar
- Open or closed watch back?
- Dial color(s)
- Print color(s)
- Color(s) of luminous elements
- Strap/Bracelet material(s)
- Strap/Bracelet color(s)
- Type and width of closing system

(b) **Organizational Briefing**

In addition to the technical briefing, the organizational briefing provides important information for the designer. It should include the following points:

- Project priority
- Date of first design evaluation
- Date of final design evaluation ("Design Freeze")
- Date of product launch
- Product manager in charge of the project
- Constructor(s) in charge of project
- Special requirements for the engraving (naming, limitation, …)

3.5 Market and History Research Phase

Market analysis is a critical step in the design of any new product. Knowledge of relevant competitors and competing products is essential for exploring market design quality and reducing the risk of unknowingly creating plagiarism. As a result of the analysis, the designer should create the following boards:

- Analysis of comparable products offered by competitors (board with a selection of products)
- Analysis of the target group and its lifestyle (mood board)
- Analysis of actual trends (board showing pictures of trendy colors, shapes, materials)
- Gathering inspiration from iconic watches of the commissioning company (board with pictures of brands' historic products)

A *mood board* is a visual tool used in the design process to visualize moods and design directions. During the design process, the mood board serves as a reference point to ensure that the design aligns with the original vision.

Fig. 3.3 Mood board with design icons of the 1960s/1970s

A mood board may be a panel of architectural cardboard with images, color, and material samples glued to it. These images, etc., can reveal characteristics of the environment that the people targeted in the design prefer. The purpose of the mood board is to gain a sense of the esthetic preferences of the target audience [90].

Figure 3.3 shows a mood board intended to reflect the design and fashion style of the 1960s/1970s. It was used as a stylistic guideline for the watch shown on the following pages.

3.6 Exploration of Brands' Design Codes

Manufacturers with a long tradition often maintain a very extensive collection of historic watch models from their own brand. Some companies, such as the International Watch Co. in Schaffhausen, even have their own watch museum. Here, inspiration for new models can be found, as well as information about company-specific product features.

The elements of a company's *product design DNA* can lie in several key areas that consistently define and distinguish its products. These elements typically include:

Form and Sign Language

- Characteristic shapes, lines, fonts, and proportions that give products a recognizable silhouette or face.

Material Choice

- Consistent use of specific materials that reflect the brand's values, such as premium quality.

Color Palette

- A distinct set of colors regularly used in product lines, often aligned with the brand identity.

Surface Treatment and Textures

- Specific finishes like matte, gloss, brushed metal, or soft-touch coatings that influence tactile perception.

User Interface and Interaction Design

- Signature ways of interacting with a product, such as button placement.

Brand Symbolism or Logos

- Consistent logo placement, size, or integration into the product without disrupting design esthetics.

Functionality Integration

- A philosophy about how to integrate features seamlessly into the design (minimalism vs. high visibility of function).

Historical Influences

- Features that have contributed to brand recognition since the introduction of the iconic models in the past.

Sustainability Principles

- Design elements that reflect environmental responsibility—repairability, recyclability.

Information about these elements should be included in a company's *Corporate Design Handbook*—especially if the company wants to ensure consistency across all product designs and maintain a strong, recognizable brand identity. The product design DNA reflects the visual and emotional identity of a brand. Including it in the design handbook ensures that all teams (internal and external) understand how product design supports and reinforces the brand [91].

3.7 Gathering of Technical Information

Before beginning the design work, the dimensions of the movement intended for the project must be determined. For this purpose, in-house technicians or external suppliers provide the so-called *cage plan*. This technical document describes all the movement dimensions relevant to the designer.

If the designer is involved in selecting the movement, he must consider the following points:

- Diameter and height of the movement
- Distance of the counter hand axes from the center
- Distance of the date wheel from the center
- Positions of dial feet openings
- Possibilities for customizing the movement through special finishes
- Possibilities for using a rotor that matches the corporate design

A designer is rarely also an engineer. Therefore, he or she must consult the design engineers about specific technical specifications, such as case material thickness, depending on the required water pressure or the necessary number of case elements.

In most cases, it makes no economic sense to redesign every element of a watch. Especially when a watch is intended to become part of an established watch family, it is often possible to incorporate existing components (hands, crystals, straps, etc.). It is part of the design process to familiarize yourself with these elements before beginning the creative process:

- Length of the existing hands of the product family compatible with the chosen movement
- Geometry of existing crown, push buttons, crystals, straps, bracelets, closing systems, …
- Geometry of existing keys to open the watch back

3.8 Concept Phase

If the design brief does not contain any specifications regarding the style of the watch, the designer looks for sources of inspiration at the beginning of the concept phase. Finding inspiration is a crucial first step, and it can come from a wide range of sources—both traditional and unexpected. The following sources are possible [92]:

- Iconic products of the respective watch brand
- Current design trends in the watch industry
- Certain art historical styles (e.g., 1970s design)
- Certain architectural styles (e.g., Art Deco)
- Style elements appropriate to the type of watch (e.g., for pilots' watches, the displays from aircraft cockpits)

The designer usually sketches out initial ideas for a watch. This can be done traditionally with pencils or with ballpoint pens on paper. During the sketching phase, quantity is preferable to drawing quality. Quality develops over time through routine.

The sketching phase involves putting as many variations of a basic idea on paper as possible. Sketches, like the one shown in Fig. 3.4, help in selecting the best variation.

Once the ideal basic shape has been found, sketches, as shown in Fig. 3.5, help work out details such as the case side, the hands, the typeface, the dial printing, ...

In a further step, he translates the design that appears most suitable, as shown in Figure 3.6, into outline drawings using a fineliner.

To vary the color and material type, marker pens can be used on special paper. The outline drawings serve as the basis for these coloured marker renderings. An example is depicted in Fig. 3.7.

Sketching and colouring on a graphics tablet (also called a drawing pad) is becoming increasingly popular. Using a pen on a drawing pad simulates drawing with a pen and paper but offers more control and flexibility. Many tablets detect the pressure, tilt angle, and even the rotational movement of the pen, enabling detailed and dynamic lines. Mistakes can be quickly undone without affecting the entire sketch. Such sketches can be digitally saved, modified, and used in different projects. Sincc no physical materials are required, waste can be avoided, making the work more sustainable.

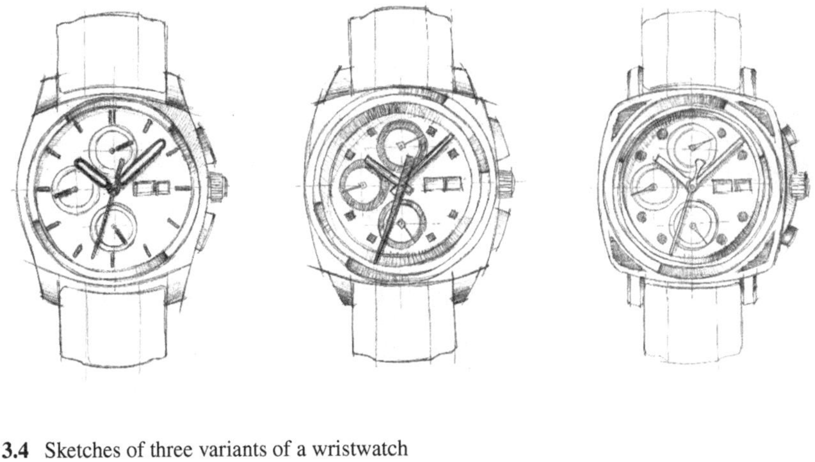

fig. 3.4 Sketches of three variants of a wristwatch

Fig. 3.5 Sketches of a wristwatch and some of its elements

3.9 Design Phase

(a) 2D-Renderings

After selecting the most suitable sketches, they are usually converted into computer renderings. If only the design differences of a dial are to be depicted, a 2D representation (see Fig. 3.8) is usually sufficient. Using a suitable program (e.g., Adobe Illustrator), these can be displayed in color with minimal time investment. Image editing programs (e.g., Adobe Photoshop) can be used to enhance the 2D renderings' realism with special effects (such as shadows, light edges, etc.) and added textures.

By saving individual elements such as hands, dial, and its printing in separate layers, their colors can be quickly changed in the programs Illustrator and Photoshop. Figure 3.9 shows color variations of the watch presented in Fig. 3.8.

Fig. 3.6 Outline drawings of a wristwatch in different perspectives

(b) **3D-Volume Models and Renderings**

3D renderings are suitable for depicting watch cases. 3D rendering programs can realistically simulate lighting, materials, and textures to achieve photorealistic results. Models can be viewed and presented from any angle, allowing for a precise representation of the design. Adjustments to color, texture, or lighting can be made quickly and easily, and multiple versions of a design can be saved and compared.

Fig. 3.7 Marker renderings of a wristwatch in different perspectives

The creation of a 3D rendering (see Fig. 3.10) takes place in several stages. Based on 2D top, front, and side drawings, a three-dimensional model is first created. This is usually done in a CAD program (e.g., Rhino). Colours, textures, or materials are then assigned to the model elements in a rendering program (e.g., KeyShot).

(c) **3D-Print Outs**

3D printers are playing an increasingly important role in the construction of watch case prototypes. 3D printers allow complex models to be created in a shorter time than with conventional production methods. Changes can be implemented quickly,

Fig. 3.8 2D rendering of a watch front

Fig. 3.9 2D colour variations of a watch front

Fig. 3.10 3D watch rendering

making the design process more agile. Unlike subtractive processes (e.g., milling), additive processes such as 3D printing use only the required material, making them more environmentally friendly. A wide range of materials, from plastics to metals, can be used depending on the requirements. The disadvantages of a 3D printer include its limited build space and large manufacturing tolerances. The layered structure of the models results in stepped surfaces that may require post-processing. While simple models can be printed quickly, detailed or large models often require several hours or even days.

The most common 3D printing methods differ mainly in how they build objects layer by layer:

FDM (Fused Deposition Modeling) melts and extrudes plastic filament through a nozzle. It is inexpensive, widely used, and suitable for prototypes, but surface quality is often moderate.

SLA (Stereolithography) uses a laser or light source to cure liquid resin into solid layers. It produces very high detail and smooth surfaces, making it popular for watch and jewelry applications.

SLS (Selective Laser Sintering) fuses powdered material—usually nylon—using a laser. It creates strong, functional parts without support structures

Metal 3D printing, such as DMLS or SLM, works similarly to SLS but uses metal powders. These methods enable the production of complex, lightweight metal components.

Binder Jetting deposits a liquid binding agent onto layers of powder. It allows fast production and can print metals, sand, or ceramics, though additional post-processing is often required.

3.10 Prototype Phase

During the prototype phase, the designer's task is to translate their designs into technical drawings. The following fundamental principles should be observed when creating a technical drawing:

1. **Clarity and Unambiguity**

   The drawing must communicate the design clearly so that every reader interprets it in the same way. All elements should be easy to understand.
2. **Accuracy and Precision**

   All dimensions, angles, radii, and proportions must be exact. A technical drawing must represent the object precisely according to its real geometry.
3. **Use of Standardized Line Types**

   Different line types must be used consistently, for example:

   - visible outlines → solid thick lines
   - hidden edges → dashed lines
   - center lines → chain lines
   - dimension lines → thin continuous lines

4. **Correct Scale**

   The drawing should be produced in an appropriate scale.
5. **Proper Dimensioning**

   Every necessary size and position must be defined by clear dimensions. Dimensions should be complete but not duplicated unnecessarily.
6. **Use of Standard Views and Projections**

   Objects should be represented using standardized views such as:

   - front view
   - top view
   - side view
   - section view
   - isometric view

7. **Compliance with Standards**

   The drawing must follow recognized standards, for example:

   - **ISO 128** (general drawing principles)

- **ISO 129** (dimensioning)
- **ASME Y14** (U.S. engineering drawings)

8. **Readable Lettering and Symbols**

   Text, symbols, tolerances, and notes must be legible and use standardized notation.

9. **Title Block and Metadata**

   Every drawing should include a title block containing:

   - drawing title
   - part number
   - scale
   - date
   - author/designer
   - revision number

10. **Tolerances and Surface Information**

    Manufacturing tolerances, fits, and surface finish requirements should be clearly specified where necessary [93, 94].

Precise plans, as the dial geometry plan illustrated in Fig. 3.11, are required to produce the first watch prototypes. The 3D design files serve as the basis for engineers to create technical drawings, which, for example, detail the geometry of the dial.

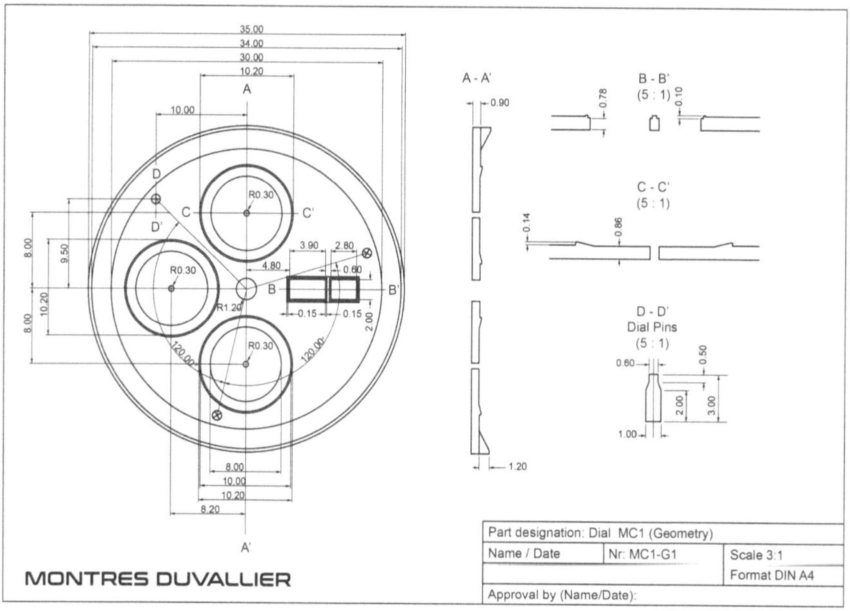

Fig. 3.11 Dial geometry plan

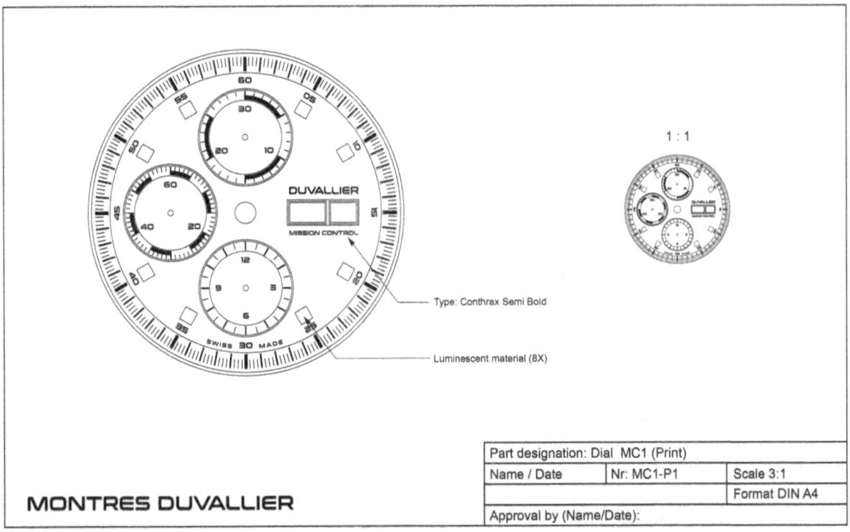

Fig. 3.12 Dial print plan

Designers supplement these drawings with plans covering printing (see Fig. 3.12), colouring (see Fig. 3.13), surface treatment (see Fig. 3.14), Back engraving pan (see Fig. 3.15) and Hands Plan (see Fig. 3.16).

Technical drawings that are issued to suppliers usually contain several legal notices and technical disclaimers. These notes protect intellectual property, limit misuse, and clarify manufacturing requirements. Although the exact wording varies by company and industry, the following categories are widely used in engineering practice

(a) **Ownership and Copyright Notice**

This statement clarifies that the drawing is the property of the issuing company and may not be copied or used without authorization.

Example wording:

- This drawing is the property of [Company Name].

  It may not be reproduced, disclosed to third parties, or used for any purpose other than that for which it was supplied without written permission.

(b) **Confidentiality Notice**

A confidentiality statement ensures that suppliers treat the information in the drawing as proprietary.

Example wording:

- This document contains confidential information.

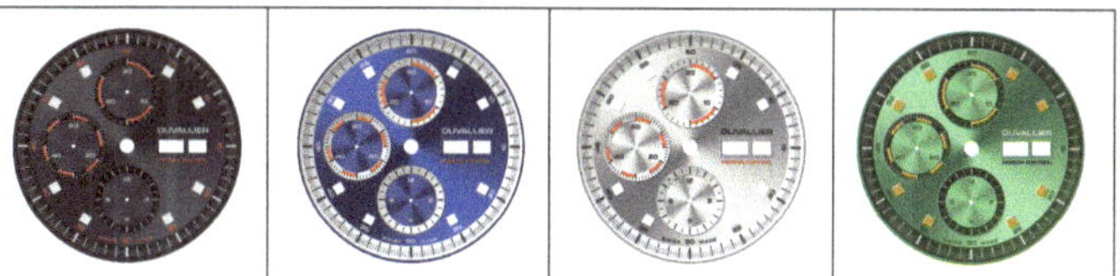

Version:	1 (Black)	2 (Blue)	3 (Silver)	4 (Green)
Dial Surface:	Solar Brush	Solar Brush	Solar Brush	Solar Brush
Dial Color:	Galvanic Black	Galvanic Blue	Galvanic Silver	Galvanic Green
Dial Print 1:	Pantone 000C White	Pantone 000C White	Pantone Black C	Pantone 3537 C
Dial Print 2:	Pantone 032C Red	Pantone 021C Orange	Pantone 032C Red	Pantone Black C
Subdial Surface:	Circular Brush	Circular Brush	Circular Brush	Circular Brush
Subdial Color:	Galvanic Black	Galvanic Blue	Galvanic Silver	Galvanic Green
Subdiall Print 1:	Pantone Black C	Pantone 000C White	Pantone 000C White	Pantone 3537 C
Subdial Print 2:	Pantone 000C White	Pantone Black C	Pantone Black C	Pantone 000C White
Subdial Print 3:	Pantone 032C Red	Pantone 021C Orange	Pantone 032C Red	Pantone 137 C
Rehaut Surface:	Painted	Painted	Painted	Painted
Rehaut Color:	Pantone Black C	Pantone 000C White	Pantone 000C White	Pantone 3537 C
Rehaut Print:	Pantone 000C White	Pantone Black C	Pantone Black C	Pantone 000C White
Lumininous Elements:	C1 White	C1 White	C1 White	FNS 1140 Orange

Part designation: Dial MC1 (Colors)		
Name / Date	Nr: MC1-C1	Scale 3:1
		Format DIN A4
Approval by (Name/Date):		

MONTRES DUVALLIER

Fig. 3.13 Dial color plan

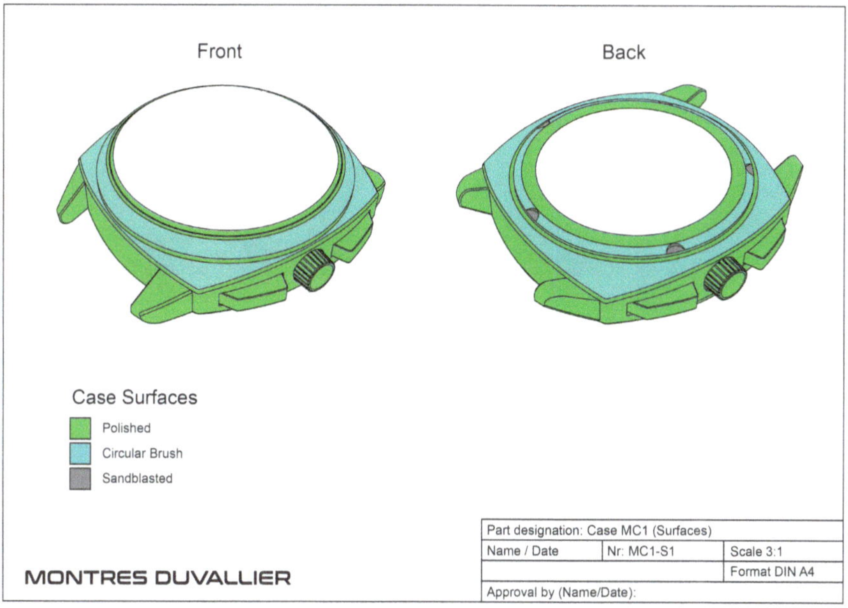

Fig. 3.14 Case surface plan

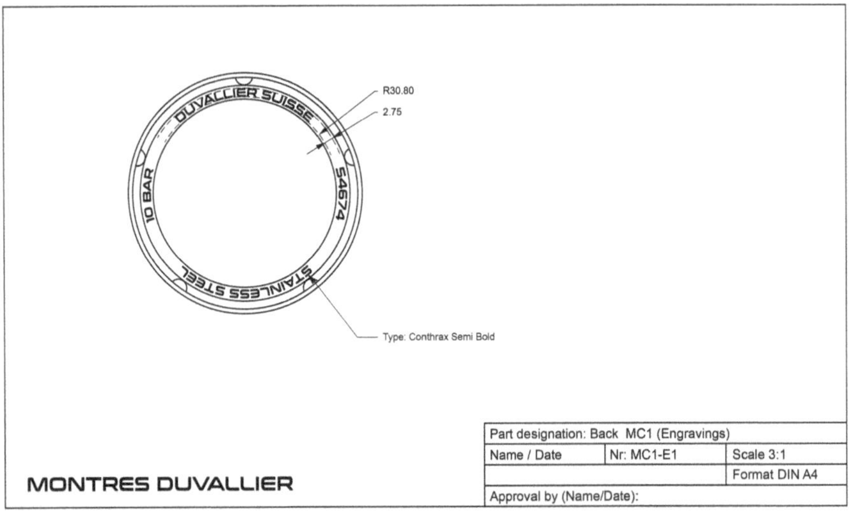

Fig. 3.15 Back engraving plan

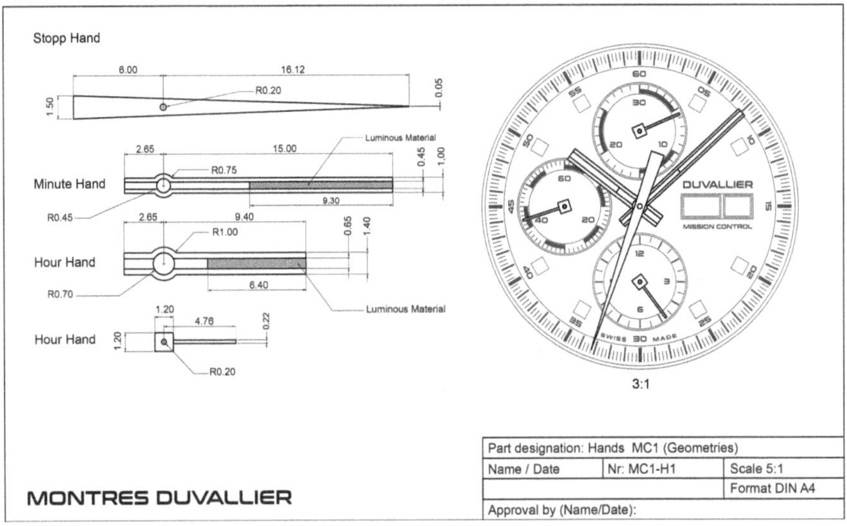

Fig. 3.16 Hands plan

It may only be used for the preparation of quotations or the manufacture of parts for [Company Name].

(c) **Restriction of Use**

A restriction clause defines the permitted purpose of the drawing.

Example wording:

- This document may only be used for manufacturing parts ordered by [Company Name].

(d) **Dimensional and Interpretation Notes**

Technical notes are essential because engineering drawings are legally binding manufacturing documents. If tolerances or units are not defined, disputes may arise.

Typical notes include:

- Do not scale the drawing.
- All dimensions in mm unless otherwise specified
- General tolerances according to ISO 2768
- Dimensions apply after surface treatment unless otherwise stated

Standards such as those from the International Organization for Standardization define rules for technical drawings and tolerances.

(e) **Dimensional and Interpretation Notes**

Each drawing should include revision information to ensure suppliers use the correct version.

Typical elements:

- Revision index (Rev A, Rev B, etc.)
- Description of change
- Date of revision
- Responsible engineer

Standards for drawing structure and title blocks are defined in ISO technical product documentation standards.

(f) **Supplier Verification or Responsibility Note**

Some companies add statements clarifying that the supplier must review the drawing before manufacturing.

Example wording:

- The supplier is responsible for verifying all dimensions and specifications before manufacturing.

(g) **Reference to Technical Standards**

Drawings often reference external standards that govern presentation and dimensioning.

Examples:

- ISO 128—general principles for technical drawings

- ISO 2768—general tolerances
- ISO 16016—protection notices for proprietary information

ISO standards define the structure, graphical representation, and documentation elements required in engineering drawings [95, 96].

Exercises 3

3.1 Create a mood board depicting the lifestyle of the following target group:

Male or female, aged between 25 and 45, high income, active in extreme sports (triathlon, surfing, skydiving, etc.) in their free time, and with a preference for sporty-casual clothing.

3.2 Use the insights from the mood board to design a sports chronograph that stylistically matches the target audience's taste. First, create several different designs as rough pencil sketches. Then, develop your favourite design into a marker rendering.

3.3 Choose a suitable movement for the chronograph you wish to design. Cage plans for chronograph movements can be found on the websites of leading movement manufacturers (e.g., www.sellita.ch). Then, create a three-dimensional model of the selected movement using a design program (e.g., Rhino).

3.4 Design the front of the chronograph in a 2D program (e.g., Adobe Illustrator). Add effects and textures in an image editing program (e.g., Adobe Photoshop). Create variations with different case and strap materials. Vary the colours of the dial and its printing, as well as the hands.

3.5 Build a 3D model of the chronograph around the selected movement. Create a coloured rendering of the watch against a neutral background.

3.6 Create the most accurate technical drawing possible of the watch dial.

Chapter 4
Tutorials

Abstract Chapter 4 presents practical tutorials for visualizing wristwatch designs throughout the design process. It covers hand sketching, emphasizing proportion, perspective, line quality, shading, and sketching as a tool for creative ideation. It then explains 2D digital rendering, including layering, lighting, materials, textures, and Photoshop post-processing. The 3D-rendering tutorial outlines case construction, integration of movement, dial, hands, strap, materials, lighting, and final rendering. Finally, the chapter addresses watch photography, distinguishing technical pack shots from atmospheric mood shots and providing guidance on composition, lighting, camera settings, reflections, and presentation.

The ability to depict watches is an essential tool for a watch designer. The following explains how to create initial hand sketches, render watches in 2D and 3D on the computer, and take successful photographs.

The use of artificial intelligence is deliberately not discussed. The use of this technology for image creation certainly has its advantages and justification. However, it carries the risk of displacing the ability to draw by hand and to design independently on the computer.

4.1 Watch Sketching Tutorial

A good hand drawing of a wristwatch is defined by the balance between technical accuracy and esthetic clarity. At its core, it must communicate the form, construction, and character of the watch in a convincing and readable way. The foundation of such a drawing lies in correct proportions and perspective. The case diameter, thickness, and lug geometry must be consistent, while the dial must appear perfectly centred and circular—or correctly elliptical when shown in perspective. Because watches are highly symmetrical objects, even small deviations in alignment or proportion are immediately noticeable.

G. Plange, *A Tutorial for Watch Designers*,
https://doi.org/10.1007/978-3-032-33722-1_4

Equally important is the quality of the line work. A strong drawing uses clean, confident, and intentional lines rather than rough or repetitive strokes. Variations in line weight help distinguish between outer contours and internal details, guide the viewer's eye, and create a sense of depth. This clarity in construction allows the viewer to understand the object quickly and intuitively.

Light and shadow play a central role in giving the drawing volume. A consistent light source allows for realistic shading across curved surfaces, such as the case and bezel. Gradual tonal transitions, combined with stronger contrasts at edges and reflections, create a convincing three-dimensional effect. Shadows, especially beneath the lugs or between applied markers and the dial, reinforce the sense of depth.

At the same time, the drawing must remain legible. Markers, hands, and any text elements should be simplified but still accurate, ensuring that the design can be easily understood without unnecessary clutter. The goal is not to reproduce every microscopic detail, but to communicate the essential features effectively.

Finally, high-quality drawings demonstrate economy and intent. They use only as many lines and as much shading as necessary, avoiding excess while ensuring that every mark serves a purpose. This restraint is a hallmark of professional product sketching.

Hand sketching is a fundamental tool for divergent thinking in design because it:

- supports rapid generation of many ideas,
- reduces cognitive load by externalizing thoughts,
- encourages exploration without commitment,
- facilitates reflective visual thinking,
- helps prevent design fixation,
- and leverages embodied cognitive processes that can stimulate creativity.

For these reasons, hand sketching remains a highly effective method for early-stage design ideation, even in an increasingly digital design environment.

On the following page, Fig. 4.1 shows sketch examples from the author's notebook.

It's relatively easy to depict a watch from the front. As shown in Fig. 4.2, you start with outlines. In a second step, hatching is added to suggest reflections and shadows (Fig. 4.2).

The intensity of the reflections allows conclusions to be drawn about the material being depicted. Glossy materials show reflections with strong light/dark contrasts. The boundaries between light areas and dark reflections are sharply defined. Matte materials, on the other hand, show no reflections. Gentle transitions into darkness indicate areas that are more in shadow than others [97].

Drawing a watch in three dimensions is considerably more challenging than sketching its front view. It is advisable to first draw rough geometric shapes (cuboids, cylinders). These basic forms are then graphically combined to create the actual shape of the clock.

Figure 4.3 shows the process from a simple cuboid to a finished sketch of a wristwatch.

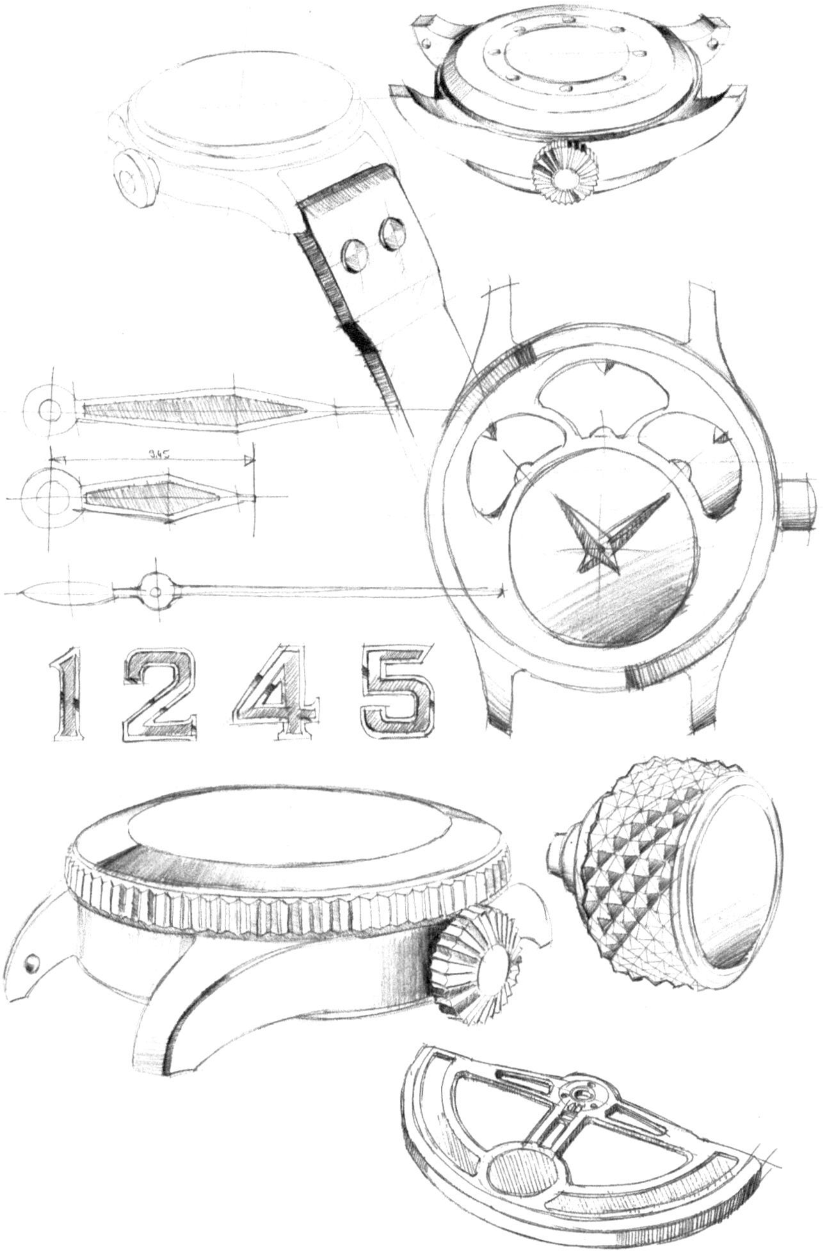

Fig. 4.1 Sketches from the author's notebook

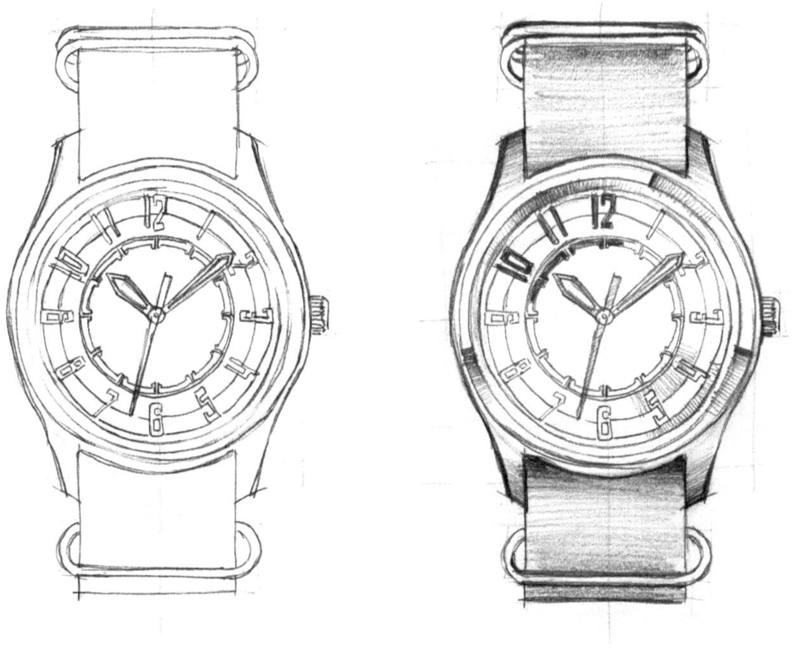

Fig. 4.2 Outlines and shaded sketch of a watch in front perspective

Ideally, the rough geometries (cuboids and cylinders) are drawn with a hard pencil that produces fine, light lines. The rest of the sketch should be done significantly darker with soft pencils. When the finished sketch is scanned, the light lines can be removed in Photoshop by increasing the contrast, as shown in Fig. 4.4.

Reversing a scanned sketch in Photoshop INTO a white image on a black background can give it a distinctive artistic touch (see Fig. 4.5).

Marker renderings of wristwatches are used to quickly and convincingly communicate a watch's shape, materials, proportions, and emotional character before a final CAD model or photorealistic rendering is produced. Successful watch marker renderings combine technical precision with expressive visualization.

The foundation of every convincing watch rendering is an accurate outline drawing done with a fineliner.

Marker renderings, as the one shown in Fig. 4.6, are usually not intended to be photorealistic paintings. Their purpose is to communicate design intent quickly and clearly. Scanned marker renderings can be further edited in Photoshop. For example, in the rendering above, a background created with markers was added, along with a shadow.

The rectangle in the background of the color-rendered watch follows the principle of overlapping. It gives the drawing more depth by representing another, more distant plane [98].

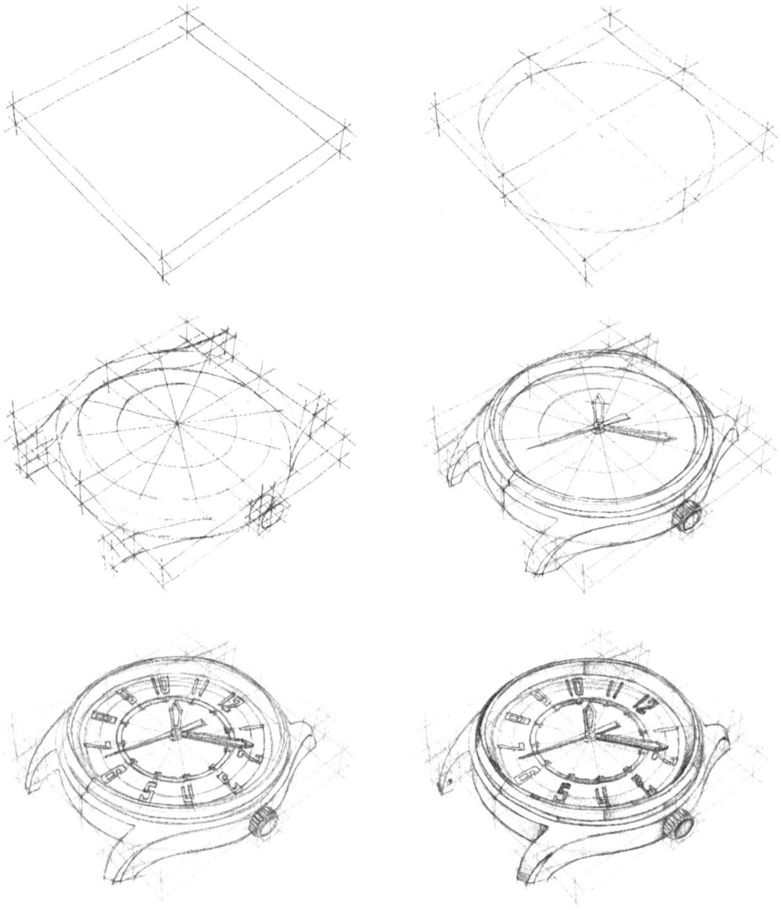

Fig. 4.3 Steps of drawing a watch 3-dimensional by hand

4.2 Watch 2D-Rendering Tutorial

Computer technology makes it possible to create two-dimensional images that appear photorealistic. The front and back of a wristwatch can be rendered quickly, making them suitable for internal presentations. Changing colors and materials can be done in just a few minutes. How this works is explained in a step-by-step tutorial.

- An outline drawing of the wristwatch to be designed is the first step in creating a 2D rendering. The following images were created using the widely used graphics program Adobe Illustrator (see Fig. 4.7).

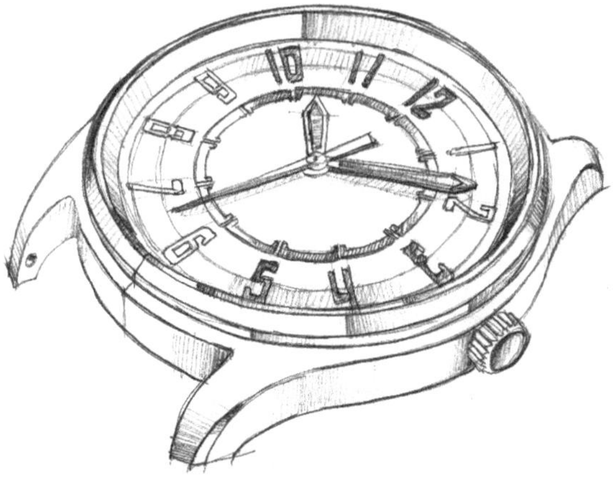

Fig. 4.4 Watch sketch reworked in Photoshop

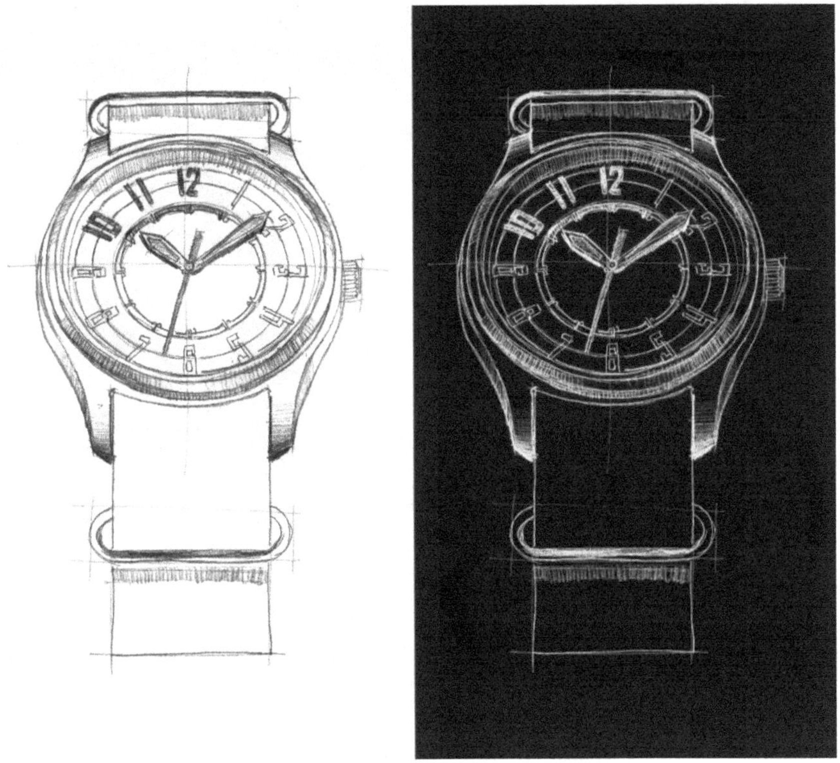

Fig. 4.5 Watch depicted with black lines on white and with white lines on black

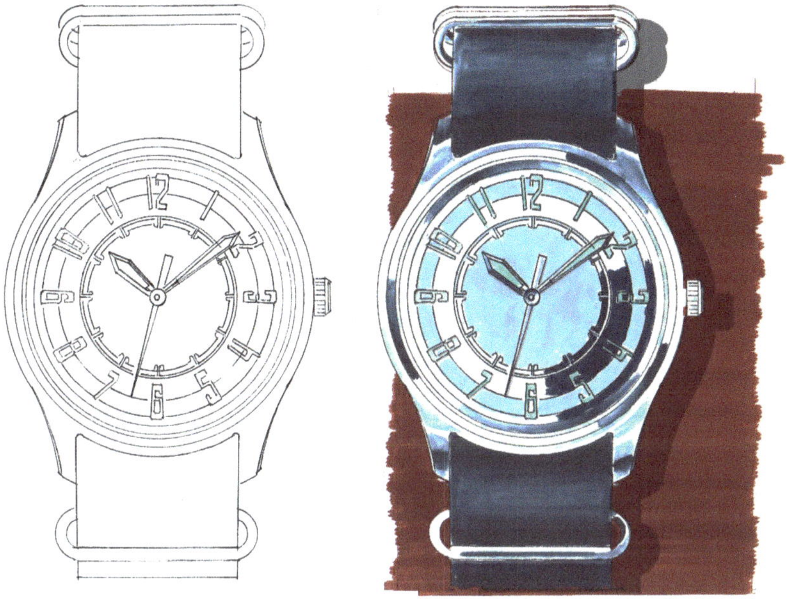

Fig. 4.6 Outline sketch and marker rendering of a watch front

- A particular challenge when creating a 2D watch rendering is the crown. The front view helps in correctly representing the side view. Guidelines can be used to transfer the crown's teeth from the front view to the side view (see Fig. 4.8).
- Related elements should be saved on separate layers. The layers should be arranged so that the ones containing elements at the top of the watch are above those containing deeper details (see Fig. 4.9).
- The elements of the line drawing are to be converted into coloured areas in the next step. Where appropriate, these areas will have a gradient of brightness. Beforehand, it must be decided from which direction the light should fall on the object to be rendered. The direction of the light determines the shadow on a watch and the shadow cast by it (see Fig. 4.10).
- When colouring surfaces, it is important to generally avoid outlines. Outlines are only useful as light edges or joints in a coloured rendering. High-gloss surfaces should be designed with reflections that create a strong contrast between black and white. Matte surfaces, on the other hand, do not exhibit reflections (see Fig. 4.11).
- For post-processing the graphic file, an image editing program such as the popular Adobe Photoshop is recommended. When exporting to a Photoshop file, ensure that the layers are preserved.

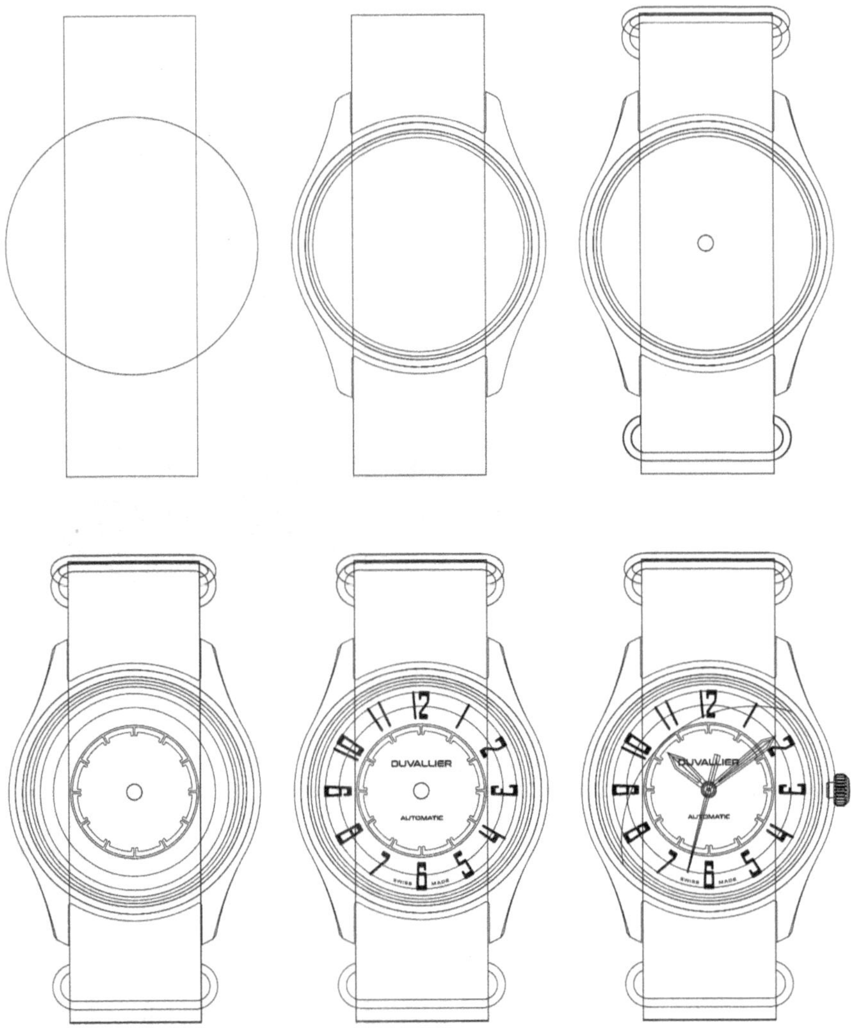

Fig. 4.7 Steps of drawing a watch on the computer

- In Photoshop, it's now possible to add shadows and round off contours. Furthermore, colors, brightness, and contrast can now be adjusted. Filters help to enhance the impression of a realistic image. Masks allow you to edit details separately (see Fig. 4.12).
- Textures, such as special metal surfaces and strap materials, can be imported into the image. For deformed surfaces like the strap, the texture must be adjusted to match its curvature (see Figs. 4.13 and 4.14).
- Perspective distortion of the band ends and a cast shadow contribute to the realistic impression (see Fig. 4.15).

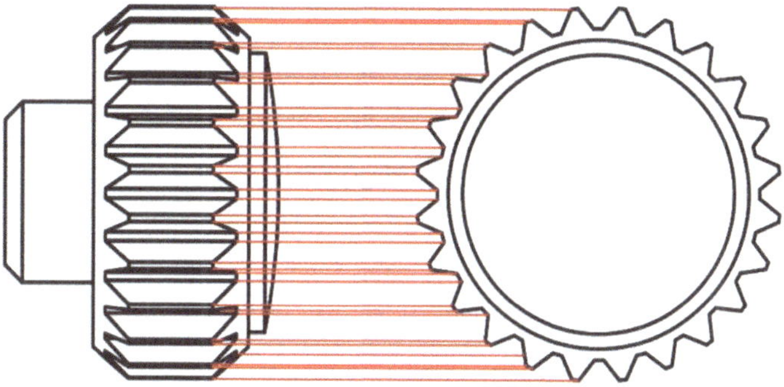

Fig. 4.8 Crown side view and front view

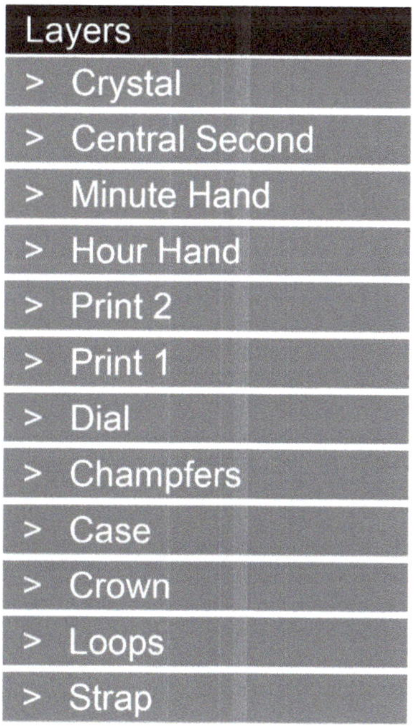

Fig. 4.9 Layers for separating the watch components

A wristwatch designer is advised to compile a collection of frequently used texture images. These textures primarily relate to the surface finishing of metallic cases, as shown in Fig. 4.16. However, it can also be worthwhile to gather leather and fabric textures for rendering watch straps.

Fig. 4.10 Direction of light in a watch 2D-rendering

Fig. 4.11 Front drawing of a wristwatch created with illustrator

4.3 Watch 3D-Rendering Tutorial

Creating a 3D model of a watch is the most important phase in the design process. This step defines the watch's geometry. The 3D file serves as the basis for the engineers' construction and as the foundation for the designers' 3D renderings.

Fig. 4.12 Front drawing of a wristwatch edited in Photoshop

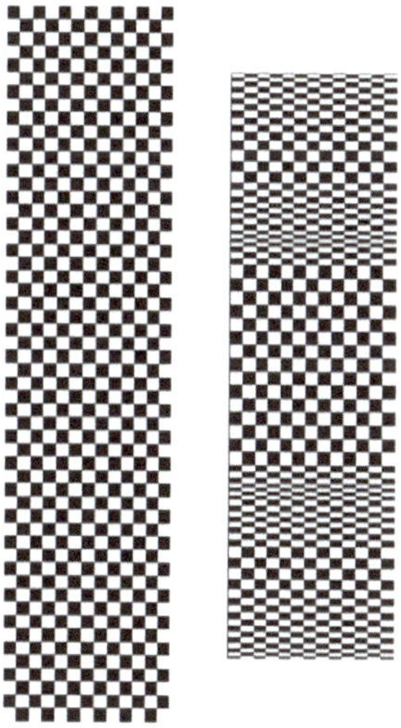

Fig. 4.13 Texture before and after distortion

- As a first step, a 3D file of the movement intended for the watch must be opened, if available. If not available, a three-dimensional model of the movement must be created based on a plan. Start your work in the XY view (see Fig. 4.17).
- If the case diameter has already been defined, it should be drawn as a circle around the movement (see Fig. 4.18).

Fig. 4.14 Front drawing of a wristwatch with texture added on strap

Fig. 4.15 Warped watch front drawing casting a shadow

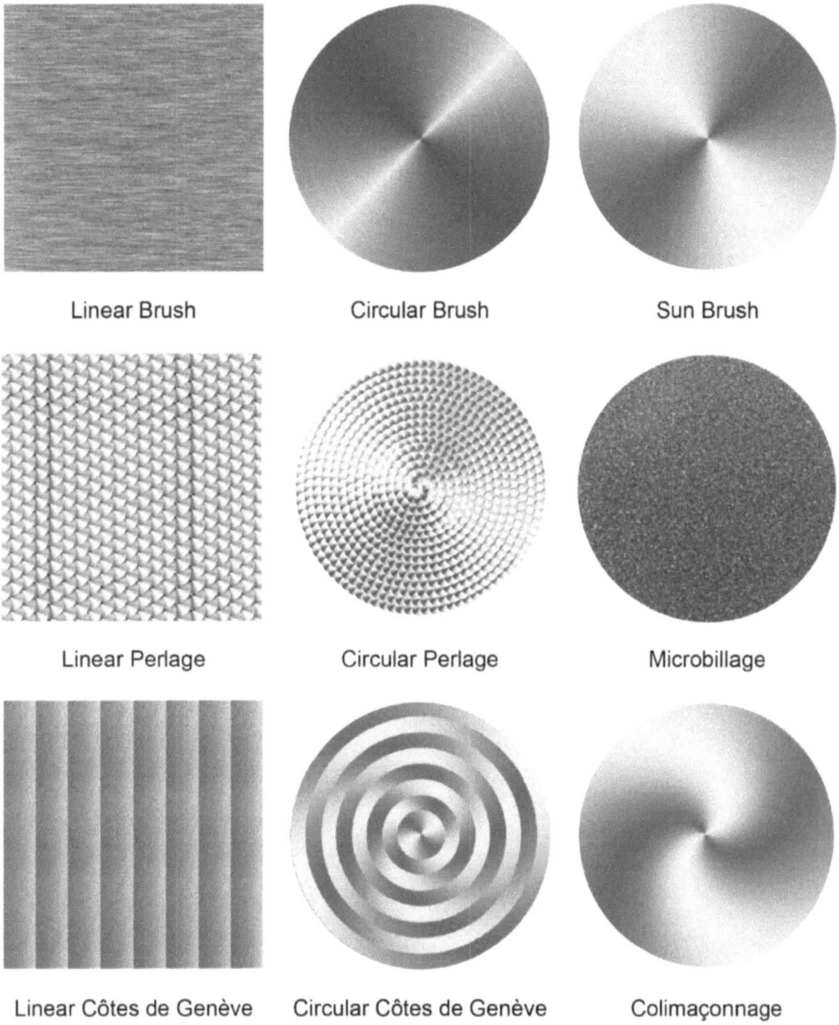

Fig. 4.16 Useful textures for watch designers

- The planned lug spacing of the watch (band width at the watch head + approximately 0.1 mm tolerance on each side) is represented as a rectangle. The transitions between the inner lug surfaces and the housing should be rounded (see Fig. 4.19).
- If the lugs are to follow a circular contour, their length must now be determined by another circle (see Fig. 4.20).
- Designing the transition from the housing to the lug comes next (see Fig. 4.21).
- Extrusion contours are created by intersecting and connecting the lines (see Fig. 4.22).

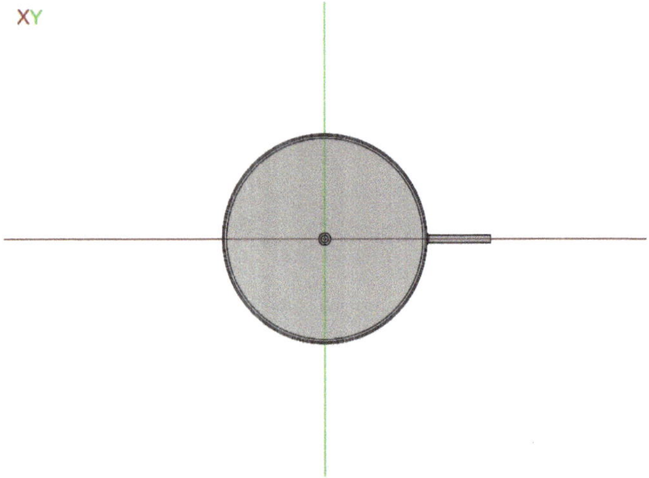

Fig. 4.17 Movement in XY view

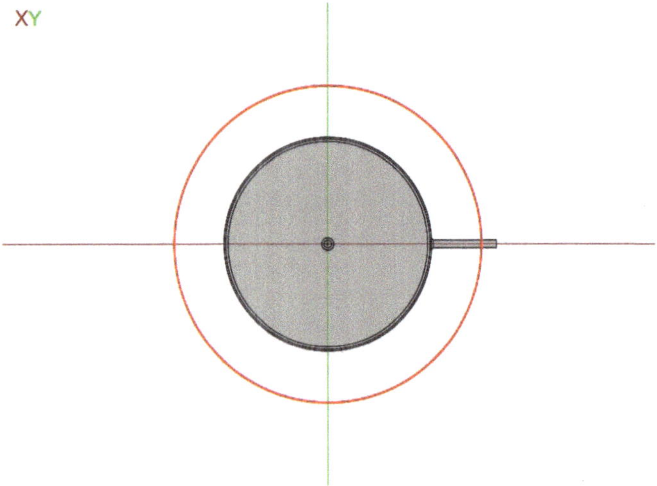

Fig. 4.18 Case diameter in XY view

- Switch to XZ view. Create a rotating contour for the case around the movement that ends at the tip of the lugs. This step requires the utmost care. When constructing the rotating contour, the necessary wall thicknesses, the dimensions of the dial, hands, crystals, gaskets, and so on must already be taken into account (see Fig. 4.23).
- Switch to XYZ view. Use the rotation contour to create a rotational object by rotating it around the z-axis (see Fig. 4.24).
- Use the extrusion contours to create extrusion surfaces (see Fig. 4.25).

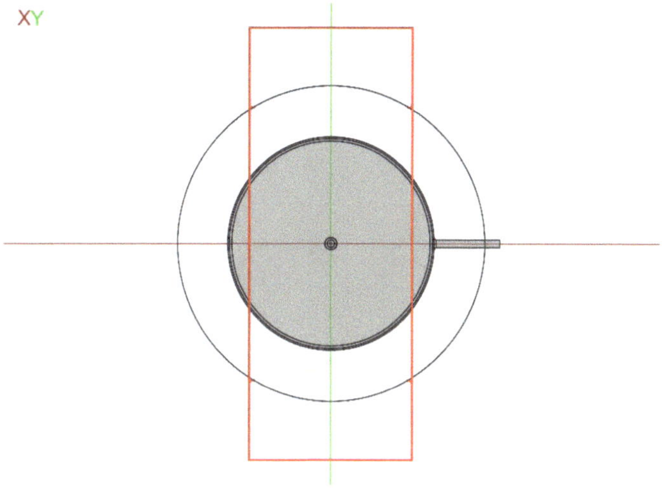

Fig. 4.19 Lug spacing in XY view

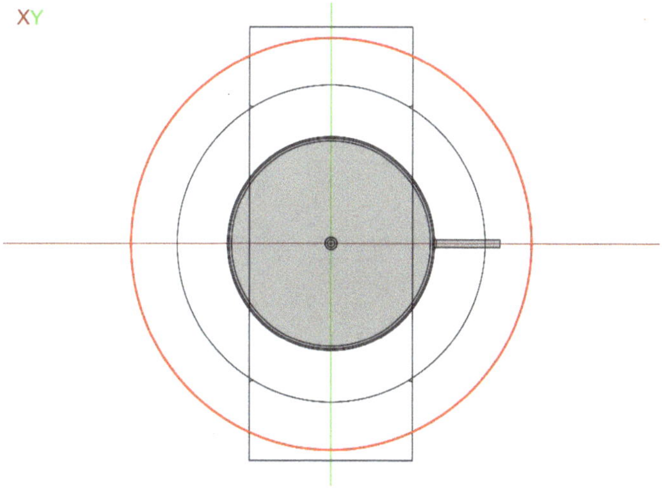

Fig. 4.20 Lug length in XY VIEW

- Intersect the rotated object with the extruded surfaces. If the result is not satisfactory, the extrusion contours and the rotation contour must be corrected (see Fig. 4.26).
- Once the case no longer requires any improvements, the lugs can be chamfered to soften them. After that, it's a good idea to add the crystal and crown (see Fig. 4.27).

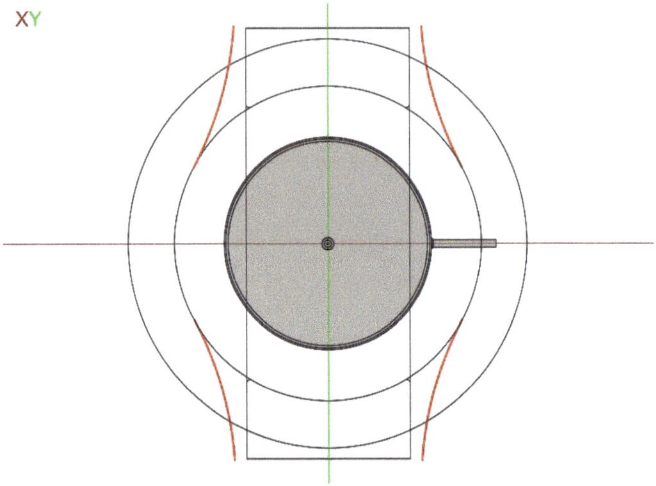

Fig. 4.21 Case transition in XY View

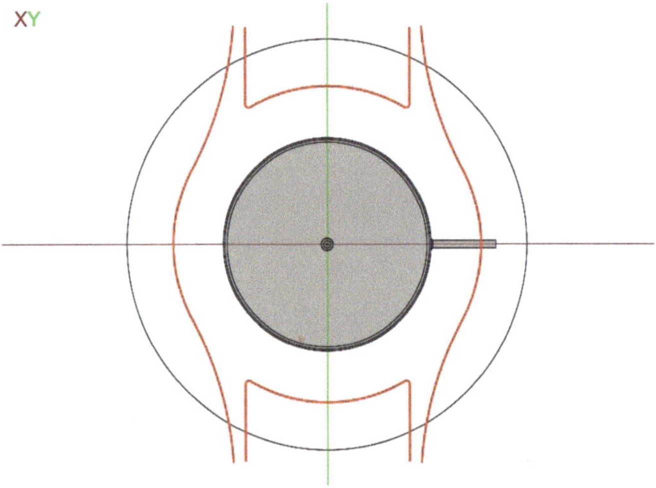

Fig. 4.22 Case extrusion contours in XY view

- Now it's time to turn the attention to the dial and hands. The graphics and lettering for the dial should be created in a drawing program such as Illustrator and then imported into the design program (see Fig. 4.28).
- The next step is to determine the details of the case back (engravings and case back key recesses). The engraving text should also first be created in a graphics program (see Fig. 4.29).

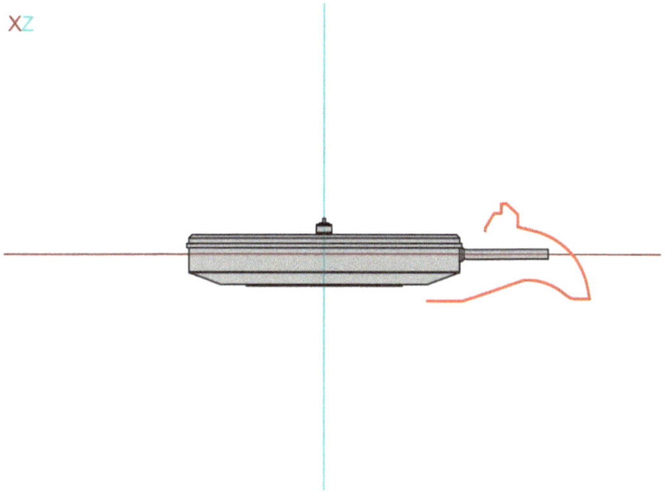

Fig. 4.23 Case rotation contour in XZ view

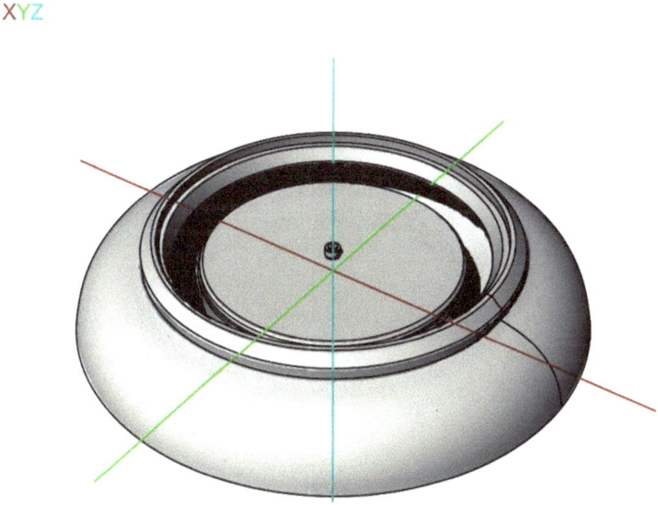

Fig. 4.24 Rotation object in isometric view

- When constructing the watch strap or bracelet, a cross-section of an average male or female wrist will help to find the correct proportions (see Fig. 4.30).
- After the watch created in a CAD program has been saved in a suitable file format, it must be opened in the rendering program. The still colourless model is called the clay model (see Fig. 4.31).

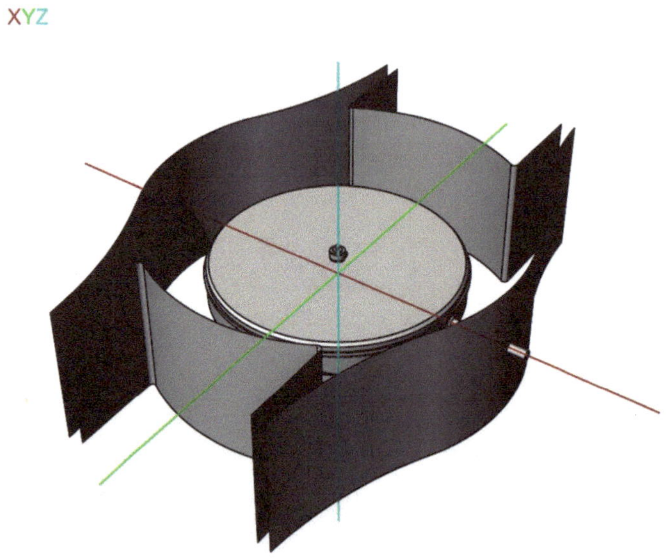

Fig. 4.25 Extrusion surfaces in isometric view

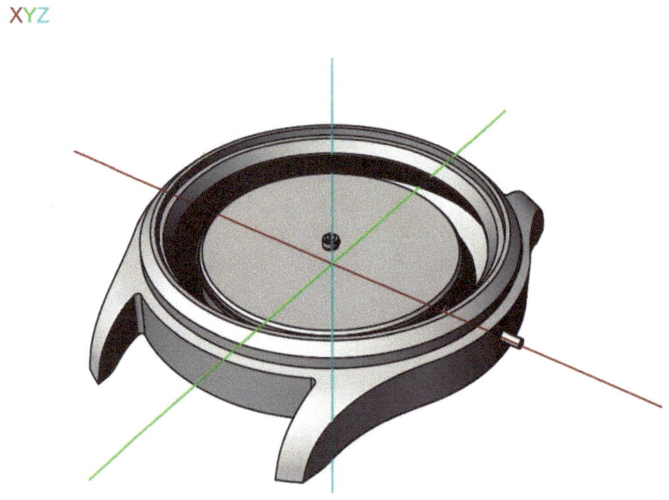

Fig. 4.26 Intersection in isometric view

- The selected materials, textures, and colors must then be assigned to the individual components of the watch. A raw rendering is created (see Fig. 4.32).
- The surroundings and lighting of the watch should be adjusted until an attractive, photorealistic image is achieved. Regarding the lighting, not only its direction, its intensity, and color, but also the quality of the shadows created should

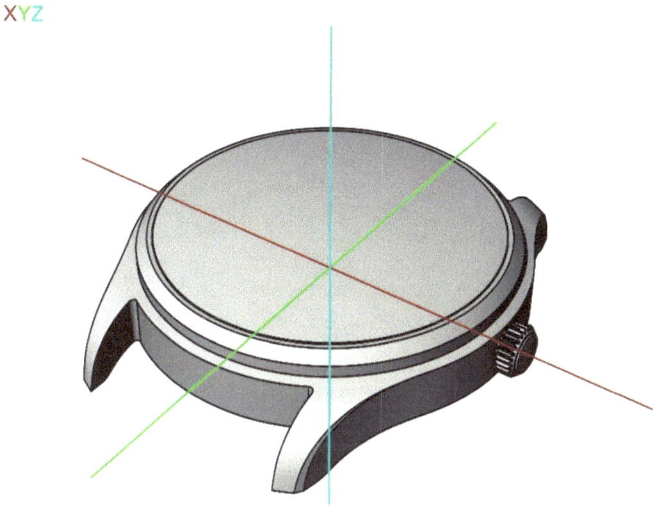

Fig. 4.27 Crystal and crown added to watch case in isometric view

Fig. 4.28 Dial and hands added to watch case in isometric view

be optimized. With glossy surfaces, it's worth experimenting with different reflections.

- Just like with a camera, professional rendering programs allow you to determine the depth of field of the rendered object. The dial and hands, as the face of a watch, should always be rendered in sharp focus, while the back of the strap can be blurred.

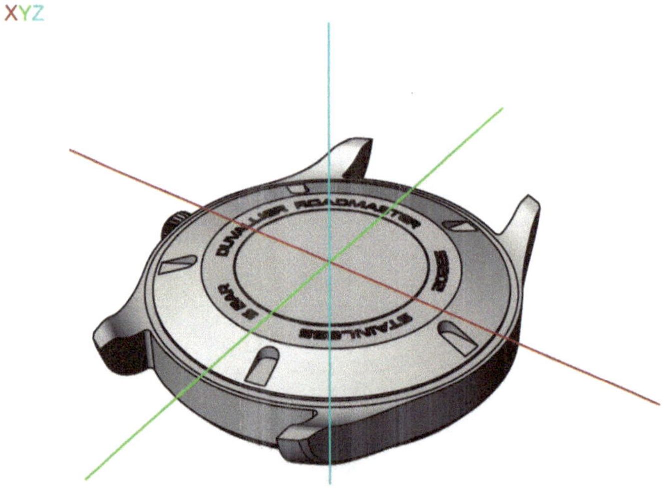

Fig. 4.29 Case back in isometric view

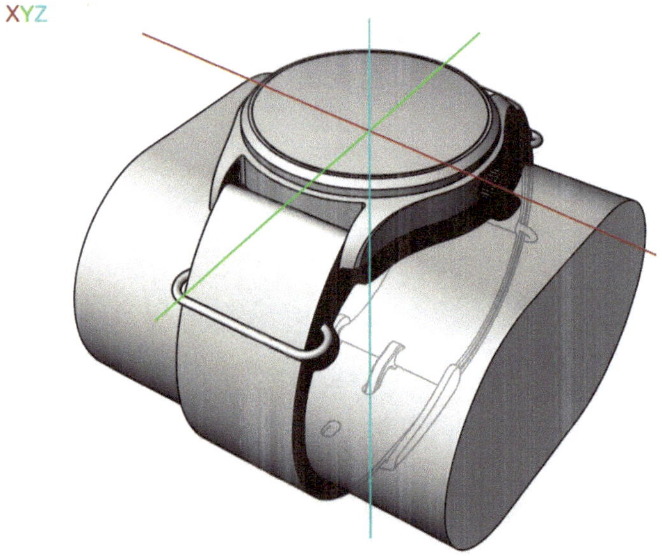

Fig. 4.30 Strap wrapped around a wrist section in isometric view

- The final rendering can be further optimized in an image editing program like Photoshop (see Fig. 4.33).

As the two examples shown in Fig. 4.34 demonstrate, watches depicted as "floating freely" also appear attractive.

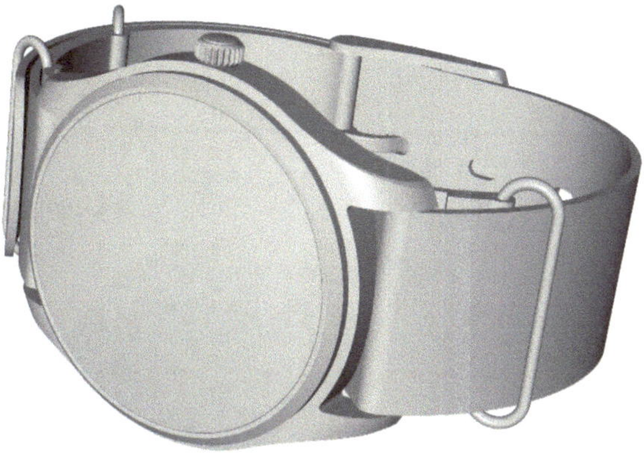

Fig. 4.31 Clay model of watch in 3D view

Fig. 4.32 Raw rendering of watch in 3D view

To visualize the technical structure of a wristwatch, creating an exploded view is a good option (see Fig. 4.35).

4.4 Watch Photography Tutorial

In the field of product photography, capturing watches is one of the most challenging disciplines. This tutorial offers tips on how to take good photos.

In a so-called pack shot (pack = packaging, shot = shot), the product is depicted as realistically as possible, either against a neutral background or in a free-standing position. Pack shots usually show a top or front view of a product to illustrate its proportions.

The following measures will help you take good pack shots of a wristwatch [99, 100]:

- First, thoroughly clean the watch. Soft microfiber cloths are ideal for this.

Fig. 4.33 Final rendering of watch with added ground in 3D view after postproduction

Fig. 4.34 3D-rendering examples of watches

Fig. 4.35 Exploded view of a watch

- Thin gloves will help you position the watch without leaving fingerprints.
- To ensure the hands are positioned correctly, set the watch to eight minutes past ten. Pull the crown just before taking the picture to prevent the hands from moving.
- When it comes to chronographs, it's worth considering whether the counter hands should be displayed in a standby position or running.
- A matte, neutral white paper is best suited as a background for the watch.
- A piece of white cardboard with a hole cut out for the camera lens in front of the watch will prevent unwanted reflections on the crystal and case.
- On highly glossy surfaces, black strips of cardboard placed next to the watch can create an interesting interplay of light and dark reflections.
- A polarizing filter helps to cut down on reflections.
- A zoom lens allows you to photograph the watch at different focal lengths.
- Focal lengths that are too short or too long will make a wristwatch appear unnatural.
- For full-frame cameras, use focal lengths between 50 and 80 mm.
- A tripod helps prevent camera shake during photography.
- Shoot in manual mode; this makes focusing easier.
- To ensure every element is in focus, use a long exposure time combined with a small aperture (high f-stop value). To achieve maximum sharpness, use the focus stacking function.

In photos of wristwatches, the hands are usually set to the 10:08 position. The hands form an upward-pointing V shape that resembles a smiling face. Advertising professionals believe that this arrangement creates subtle positive associations in viewers. On many watches, the brand logo is positioned below the 12 o'clock mark. At 10:08, the hands frame this area rather than covering it. The hands appear balanced and draw attention toward the center of the watch.

Figure 4.36 shows the pack shot of a pilot's watch. When taking the photograph, care was taken to align the wristwatch exactly parallel to the camera.

A light tent made of thin wooden sticks and gauze (see Fig. 4.37) creates soft light that will show the watch's details much better. If no light tent is available, diffuse daylight will help.

One difficulty in photographing watches is aligning them correctly from a perspective standpoint. This is especially true for watches that are to be photographed with a closed strap. A watch stand, like the one shown in Fig. 4.38, can help; it can be purchased from watch retailers or ordered online.

Unlike the sober pack shot, the mood shot (mood = mood) attempts to emphasize a product's esthetics rather than its technical features. It is shown in an environment that matches its appearance or during use. Mood shots usually present objects three-dimensionally against a background that matches the color and theme.

Fig. 4.36 Pack shot of a watch

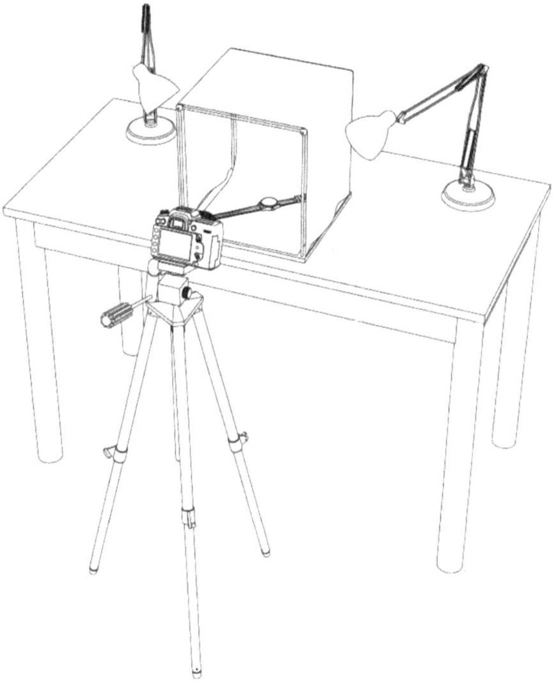

Fig. 4.37 Light tent used for watch photography

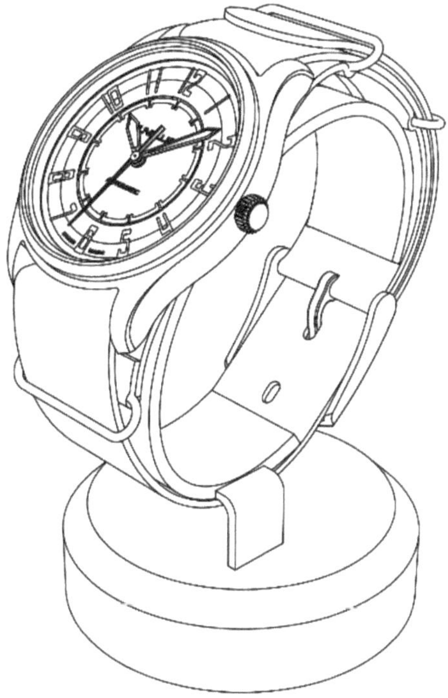

Fig. 4.38 Watch on a watch stand

A mood shot serves to showcase the product, evoking positive emotions and encouraging purchase. Creating a successful mood shot requires more time than creating a pack shot. The photographer must experiment with the effect of different camera angles on the product, different lighting situations, background colors, substrate materials, and complementary accessories until he achieves a good result [101].

Photographing a mood shot requires one thing above all: imagination! Before the shoot, consider which environment and accessories suit the character of a watch. A pilot's watch looks authentic in a suitable environment, such as an aircraft cockpit; flight plans, and similar items make excellent accessories. Water droplets on the case demonstrate the suitability of dive watches for use in the water.

The watch shown in Fig. 4.39 was carefully lit. The flight jacket chosen as a background emphasizes the watch's masculine character and hints at its (original) purpose. To prevent camera shake, the DSLR camera was mounted on a tripod, and a self-timer was used. In the Photoshop program, the sharpness, color saturation, brightness, and contrast were adjusted.

Exercises 4

4.1 Make pencil sketches of your (favourite) wristwatch. Start with a front view, then make three-dimensional sketches from different perspectives.

Fig. 4.39 Mood shot of a pilot's watch

4.2 Using a compass, a straight ruler, and a French curve, create an outline drawing of your wristwatch. Then colour the outline drawing with coloured markers. If you don't have markers, you can also use watercolours.

4.3 Create a front view of your watch on your computer using a drawing program (e.g., Adobe Illustrator). Add details such as textures, shadows, and bevels in an image-editing program (e.g., Adobe Photoshop).

4.4 If you have access to a 3D program (e.g., Rhino), try recreating your wristwatch. Then add colours, materials, and textures to the basic model. Find a suitable environment and lighting for the final rendering.

4.5 Use a camera to take a successful pack shot and a mood shot of your wristwatch. Experiment with different combinations of aperture and shutter speed, various lens focal lengths, different light source positions, and perspectives when taking the pack shot. For the mood shot, choose a setting and accessories that complement the watch.

References

1. N.A.A. (2025) Parts of a watch 101—ultimate intro guide to everything watches inside & out. https://bespokeunit.com/watches/watch-parts-guide/
2. Wilson B (2025) The 8 different types of watch movements. https://www.watchdaydate.com/archives/19397
3. Arthur B (2025) Parts of a mechanical watch movement—an illustrated guide. https://benswatchclub.com/blog/parts-of-a-mechanical-watch
4. Brunner GL, Pfeiffer-Belli C (1999) Wristwatches. Cologne, p 9
5. Brunner GL, Pfeiffer-Belli C (1999) Wristwatches. Cologne, p 12
6. N.N.A. (2026) Bulova accutron: the revolutionary tuning fork watch that changed timekeeping. https://www.strapcode.com/blogs/watch-movements-and-parts/the-revolutionary-bulova-accutron-watch-with-a-tuning-fork-mechanism
7. Woodford C (2026) Quartz clocks and watches. https://www.explainthatstuff.com/quartzclockwatch.html
8. Potter SJ (2025) A guide to watch complications. https://www.watches-of-switzerland.co.uk/calibre/inspiration/a-guide-to-watch-complications
9. Friesenegger S (2023) 101 Dinge, die man über Armbanduhren wissen muss. München, p 137
10. Reymondin C et al (2003) The theory of horology. Le Sentier, p 352
11. Reymondin C et al (2003) The theory of horology, pp 244–252
12. Reymondin C et al (2003) The theory of horology. Le Sentier, pp 30–31
13. Friesenegger S (2023) 101 Dinge, die man über Armbanduhren wissen muss. München, p 122
14. Friesenegger S (2023) 101 Dinge, die man über Armbanduhren wissen muss. München, p 120
15. Reymondin C et al (2003) The theory of horology. Le Sentier, p 204
16. Reymondin C et al (2003) The theory of horology. Le Sentier, pp 207–218
17. Friesenegger S (2023) 101 Dinge, die man über Armbanduhren wissen muss. München, pp 108–109
18. Reymondin C et al (2003) The theory of horology. Le Sentier, pp 167–169
19. Canonizado M (2026) The art of complications—tourbillons. https://www.watchadvice.com.au/28790/the-art-of-complications-tourbillons/
20. Kessler S (2026) The history and modern evolution of the original president's watch, the Vulcain cricket. https://oracleoftime.com/the-vulcain-cricket/
21. IWC Schaffhausen (2011) Portuguese Sidérale Scafusia. Schaffhausen
22. N.A.A. (2025) 30 types of watches: a complete guide. https://watchresearcher.com/types-of-watches/

G. Plange, *A Tutorial for Watch Designers*,
https://doi.org/10.1007/978-3-032-33722-1

23. N.A.A. (2026) 11 different types of watches you need to know about. https://chronohunter.com/chronicles/11-different-watch-types
24. Friesenegger S (2023) 101 Dinge, die man über Armbanduhren wissen muss. München, pp 84–85
25. Friesenegger S (2023) 101 Dinge, die man über Armbanduhren wissen muss. München, pp 94–95
26. N.A.A. (2025) Understanding the differences between men's and women's watches. https://blog.luxehouze.com/en/understanding-the-differences-between-mens-and-womens-watches/
27. Bröhan E, Berg T (1994) Avantgarde design. Cologne, p 7
28. Lindlinger H, Huchthausen C (1978) Geschichte des Industriedesigns. Hannover, pp 10–20
29. N.A.A. (2025) The evolution of watch design: a journey through the decades. https://www.lu-ann.com/post/the-evolution-of-watch-design-a-journey-through-the-decades
30. N.A.A. (2025) The fascinating history of wristwatches: from battlefield tool to status symbol. https://handwatch.com/history-of-wristwatches-evolution-timeline/
31. Pulvirent S (2026) Gérald Genta: the most visionary watch designer of the 20th century. https://www.sothebys.com/en/articles/gerald-genta-the-most-visionary-watch-designer-of-the-20th-century
32. Voyles A (2026) The world of watch designers: Jorg Hysek. https://www.chrono24.com/magazine/the-world-of-watch-designers-jorg-hysek-p_169726/
33. N.A.A. (2026) Max bill wristwatch: the design revolution genesis. https://encyclopedia.design/2024/06/18/max-bill-wristwatch-the-design-revolution-genesis/
34. Hawes C (2026) The greatest watch designers you've never heard of. https://www.watchfinder.ch/articles/feature-the-greatest-watch-designers-youve-never-heard-of?langcode=en-ch
35. N.A.A. (2026) Smart luxury: the definitive watch trends forecast. https://breakingac.com/news/2025/dec/03/smart-luxury-the-definitive-watch-trends-forecast-for-2026
36. N.A.A. (2026) Top 5 watch trends for 2025 you need to know. https://www.wixez.com/blogs/insights/top-5-watch-trends-for-2025-you-need-to-know
37. Guaccio BM (2026) Bevorstehende Uhrentrends im Jahr 2025. https://debonarwatches.com/de/blogs/watches/upcoming-watch-trends-to-watch-in-2025
38. N.A.A. (2026) Are you ready to join us? https://www.creative-academy.com
39. Connor A (2025) A guide to watch case shapes and designs. https://www.gearpatrol.com/watches/a486351/watch-case-shapes-guide/
40. van Straaten T (2026) Back to basics: watch case materials explained—from aluminium to zirconium dioxide. https://www.fratellowatches.com/back-to-basics-watch-case-materials-explained-from-aluminum-to-zirconium-dioxide/
41. Nisar U (2026) A beginner's guide to watch case finishes. https://www.hewore.com/a-beginners-guide-to-watch-case-finishes/
42. N.A.A. (2026) The IWC anti-magnetic soft iron-cage. https://www.watchprosite.com/iwc/the-iwc-antimagnetic-soft-iron-cage/727.383432.2309262/
43. Kalra R (2025) The secret to good watch design: the art of proportion. https://quillandpad.com/2023/12/02/the-secret-to-good-watch-design-the-art-of-proportion/
44. Kumar N (2026) Why watch size is important and how to find your watch size? https://fashionsuggest.in/finding-suitable-watch-for-your-wrist-size/
45. Reymondin C et al (2003) The theory of horology. Le Sentier, pp 263–264
46. N.A.A. (2025) Welches ist das beste Uhrenglas? https://www.aurifo.com/de/welches-ist-das-beste-uhrenglas
47. Belson T (2025) Types of watch backs: a complete guide to watch casebacks. https://www.jomashop.com/blog/guides/your-comprehensive-guide-to-watch-casebacks. Was ist eine Uhrengravur?
48. Maz P (2026) Level up your watchmaking game: a beginner's guide to case back presses. https://watchmakingtools.com/beginners-guide-to-case-back-press
49. N.A.A. (2025) Was ist eine Uhrengravur? https://www.aurifo.com/de/was-ist-eine-uhrengravur

50. N.A.A. (2025) The art of selecting and setting stones. https://www.hautehorlogerie.org/en/watches-and-culture/watchmaking-knowledge/encyclopedia/the-art-of-selecting-and-setting-stones
51. Svaboe T (2025) 7 styles of watch crowns you should know. https://oracleoftime.com/styles-of-watch-crowns/
52. Greiner C (2025) The watch dial: the face of fashion and functionality. https://jestikcollection.com/blogs/articles/watch-dial/
53. N.A.A. (2025) What Is the dial on a watch? types, functions & aesthetic appeal explained. https://watch-masters.com/what-is-watch-dial/
54. Scholz F (2026) The balance of symmetry and asymmetry in dial design. https://www.acollectedman.com/blogs/journal/symmetry-asymmetry-dial-design
55. Emami A (2014) 360° industrial design. Sulgen, pp 72–76
56. DeVane A (2025) Watch dial colors: meanings and popularity. https://www.exquisitetimepieces.com/blog/watch-dial-colors-meanings-and-popularity/
57. Stinson L (2025) The case for better watch typography. https://www.hodinkee.com/articles/the-case-for-better-watch-typography
58. N.A.A. (2025) Different types of watch indices. https://www.timepieceking.info/different-types-of-watch-indices/
59. N.A.A. (2026) Why watch dials glow: a guide to luminescent materials in horology. https://watchstyler.com/blogs/news/why-watch-dials-glow-a-guide-to-luminescent-materials-in-horology
60. Siems M (2026) A watch guide to numerals—types. Layouts & Evolutions. https://goldammer.me/blogs/articles/guide-watch-numerals-layout
61. Simon C (2026) Panerai California dials and the Rolex connection. https://www.bobswatches.com/rolex-blog/panerai/panerai-california-dials-rolex.html
62. Callan A (2025) The chronograph guide: types of scale. https://beckertime.com/blog/the-chronograph-guide-types-of-scale/
63. Reymondin C et al (2003) The theory of horology. Le Sentier, p 270
64. N.A.A. (2025) Glossary of watch terms/hands. https://barringtonwatchwinders.com/en-eu/pages/hands
65. Arthur B (2025) Types of watch hands—a comprehensive guide to the art of timekeeping. https://www.benswatchclub.com/blog/types-of-watch-hands
66. N.A.A. (2025) Types of leather watch straps: your complete guide. https://thrifty-gentleman.com/blogs/blog/types-of-leather-watch-straps-your-complete-guide/
67. N.A.A. (2026) Are alligator watch straps legal? full 2026 guide. https://www.beltley.com/blogs/blog/are-alligator-watch-straps-legal?
68. Rose B (2025) The best watch strap materials: a comprehensive guide to style, durability, and comfort. https://wisstraps.com/blogs/watch-blogs/the-best-watch-strap-materials-a-comprehensive-guide-to-style-durability-and-comfort/
69. N.A.A. (2025) Comparing watch strap materials: Canvas, Nylon, and FKM. https://timetrek.shop/blogs/news/comparing-watch-strap-materials-canvas-nylon-and-fkm/
70. N.A.A. (2026) The fascinating and humble history of the NATO watch strap. https://www.gearpatrol.com/watches/a85507/nato-strap-history/
71. Agar-Hutton S (2025) Which metal watch band is right for you? https://www.watchgecko.com/blogs/magazine/which-is-best-metal-watch-band
72. N.A.A. (2025) Anleitung zu jeder Art von Uhrenverschluss und Verschlussmechanismus. https://watchello.com/de/pages/guide-of-every-type-of-closure
73. Baldassarre T (2025) A guide to the various types of closures. https://teddybaldassarre.com/en-ch/blogs/watches/watch-clasps
74. N.A.A. (2026) Spring bars & quick-release watch straps—complete guide. https://dupe.watch/guides/spring-bars-quick-release?
75. N.A.A. (2026) The role of the bridge in luxury mechanical watches. https://www.reservoir-watch.com/services/glossary/bridge-luxury-mechanical-watches-explained/

76. N.A.A. (2026) Three-quarter plate. https://www.alange-soehne.com/us-en/manufacture/art-of-watchmaking/three-quarter-plate
77. N.A.A. (2026) What is Perlage? https://barringtonwatchwinders.com/pages/perlage
78. N.A.A. (2026) Finishing. https://www.crownwatchblog.com/glossary/finishing
79. N.N.A. (2026) Watch decorations and finishings. https://decovi.ch/en/service/decorations-and-finishings/
80. Reymondin C et al (2003) The theory of horology. Le Sentier, p 272
81. N.A.A. (2026) Swiss excellence certified. https://www.cosc.swiss
82. Lancaster A (2026) Poinçon de Genève, new hallmarking system. https://timetransformed.com/2014/09/21/poincon-de-geneve-new-hallmarking-system/
83. N.A.A. (2025) A deep dive into the rotor in automatic watches. https://www.hourstriker.com/learn/a-deep-dive-into-the-rotor-in-automatic-watches
84. N.A.A. (2026) The rise of watch customization: luxury brands offer bespoke creations. https://www.watchthisx.com/insights/the-rise-of-watch-customization-luxury-brands-offer-bespoke-creations
85. N.A.A. (2025) How dial legibility impacts watch functionality. https://vaswatch.com/blog/f/how-dial-legibility-impacts-watch-functionality/
86. Mizalok V (2025) Was ist Redundanz. https://miszalok.de/Lectures/L01_Redundancy/Redundancy_d.htm
87. Sabater V (2025) Das kollektive Unbewusste nach Carl Gustav Jung. https://gedankenwelt.de/das-kollektive-unbewusste-nach-carl-gustav-jung/
88. Plange G (2026) Industriedesign—Der Weg zum guten Produkt. Wiesbaden, p 22
89. Plange G (2026) Industriedesign—Der Weg zum guten Produkt. Wiesbaden, pp 23–26
90. Plange G (2026) Industriedesign—Der Weg zum guten Produkt. Wiesbaden, pp 27–28
91. Brüggemann D (2025) Corporate design manual. https://dariabruegemann.de/corporate-design-manual/
92. Lai A (2025) From concept to creation: the intricate art of designing watches. https://albertlaimd.com/watch-design-process/
93. N.A.A. (2026) Technical drawing principles and standards. https://studyrocket.co.uk/revision/level-3-construction-and-the-built-environment-btec/graphical-detailing/technical-drawing-principles-and-standards
94. N.A.A. (2026) Fundamentals of technical drawings. https://sabner.com/fundamentals-of-technical-drawings
95. N.A.A. (2026) Confidentiality; title to drawings and specifications sample clauses. https://www.lawinsider.com/clause/confidentiality-title-to-drawings-and-specifications
96. N.A.A. (2026) Legal requirements for engineering drawings: a complete guide. https://statonix.com/legal-requirements-for-engineering-drawin
97. N.A.A. (2026) How to draw a watch. https://itizenside.com/entertainment/how-to-draw-a-watch/
98. Eissen K, Steur R (2007) Sketching. Amsterdam, p 57
99. N.A.A.: How to photograph watches: a practical guide for collectors (2025). https://handwatch.com/how-to-photograph-watches-practical-guide/
100. Ferenczi B (2025) How to photograph a watch. https://www.chrono24.com/magazine/how-to-photograph-a-watch-p_18082/
101. Plange G (2024) Fotografieren—Der Weg zum guten Bild. Munich, p 212

Index

G. Plange, *A Tutorial for Watch Designers*,
https://doi.org/10.1007/978-3-032-33722-1